America's Navy and Marine Corps Airplanes

America's Navy
and
Marine Corps Airplanes
1918-to the Present

Francis H. Dean

Schiffer Military History
Atglen, PA

Dedication

This book is dedicated to Sam, Fred, Big, Alf, Chris, and Annie

Book Design by Ian Robertson.

Copyright © 1999 by Francis H. Dean.
Library of Congress Catalog Number: 98-88429

Printed in China.
ISBN: 0-7643-0557-3

We are interested in hearing from authors with book ideas on related topics.

Published by Schiffer Publishing Ltd.
4880 Lower Valley Road
Atglen, PA 19310
Phone: (610) 593-1777
FAX: (610) 593-2002
E-mail: Schifferbk@aol.com
Please visit our web site catalog at www.schifferbooks.com
or write for a free catalog.
This book may be purchased from the publisher.
Please include $3.95 postage.
Try your bookstore first.

CONTENTS

AIRCRAFT CLASSES AND MANUFACTURER'S LETTERS

(The various aircraft classes are shown in CONTENTS)

MANUFACTURER'S LETTERS ASSIGNED BY THE NAVY (1922-1962)
A Aeromarine, Atlantic/Fokker, Brewster
B Boeing, Beech, Budd
C Curtiss, Cessna
D Douglas, McDonnell
DW Dayton Wright
E Bellanca, Edo, Piper
F Grumman
G Bell, Eberhardt, Goodyear, Great Lakes
H Hall Aluminum, Howard, Huff Daland, Stearman Hammond
J Berliner Joyce, General Aviation
K Fairchild, Kaiser, Keystone, Kinner
L Loening, Bell, Columbia
M Martin, General Motors/Eastern
N Naval Aircraft Factory
O Lockheed, Piper, Viking
P Spartan
Q Fairchild, Stinson
R Ford, Ryan
S Stout, Stearman
T New Standard, Northrop, Timm, Temco
U Vought
V Lockheed, Canadian Vickers, Vultee
W Wright, Waco
Y Consolidated/Convair
Z Pennsylvania Aircraft Syndicate

PHOTO SOURCES/CREDITS

PHOTO SOURCES
AAHS-American Aviation Historical Society
    Harold Andrews
    Roger F. Besecker
    Warren Bodie
    Henry V. Borst
    Peter M. Bowers
    Martin R. Copp
Curtiss-Curtiss Propeller Division
Court-Court Commercial Photos
    Jon Davis
    Francis H. Dean (Author)
    Robert G, Dean
    Arthur J. Di Stasi
    Robert Dorr
Edo- Edo Corporation
    Robert Esposito
    William Larkins
    Andre La Clair
    Howard Levy
    Charles Mandrake
    H. Martin
    David Menard

Mfr.-Manufacturer
NACA-National Advisory Committee for Aeronautics
    Sam Orr
    Frank Schertzer
    A. U. Schmidt
    Al Schoeni
    Milton Sheppard
    Robert Stuckey
    F. Sullivan
    W. Swisher
    John J. Schneider
    N. Taylor
    Harry Thorell
USAF-United States Air Force
USCG-United States Coast Guard
USMC United States Marine Corps
USN- United States Navy
(UNK.)-Unknown
    Walsh
    Joseph Weathers
    Gordon Williams
Wright-Wright Aeronautical Corporation

A Note About Photo Credits

Many of the photos in this book are those of the airplane manufacturers; many others are Navy photos and there are a few from other US services and government organizations.

A large number of photos have also come from private sources and are often difficult to track for proper credit since many private photos have been traded over the years, and the original photographers may not be known. The author is approaching this situation a little differently than previously such that any photo, or negative providing a photo that has been given or traded into his possession has been credited to the person who traded it to him, or, if known, credited to the previous holder whether or not that person actually took the picture. The idea is to go back as far as possible towards the original photographer at least to the extent of the author's knowlege.

Many photos or negatives were obtained or borrowed from long time airplane photographers and/or photo collectors such as Hal Andrews, Roger Besecker, Pete Bowers, Chuck Mandrake, Sam Orr, and Joe Weathers.

Some photos are credited to Harry Thorell; these were obtained by the author in the period 1939 to about 1943 from the Craft Art Company in California. Apparently not all the Craft Art photos are from the Harry Thorell collection; others may have had a hand in Craft Art, and some of the photos may well be copies of government or manufacturer's shots, but in any case Harry Thorell has been cited as the source.

INTRODUCTION

This book is about US Navy airplanes and thus also covers photographically most planes used by the US Marines and the US Coast Guard. It is not a history of Naval Aviation but rather a photo essay showing the changes in Navy airplane designs over the years. The period covered is roughly from 1918, the end of World War I, up to the present time, though a few earlier airplanes are shown. Most all Navy types are included, but a few relatively obscure models are not. This fact should bother only the most studious of naval aircraft historians.

The book is mainly pictorial and is organized in sections covering the various types of Navy aircraft. The sections are arranged in alphabetical order by airplane type starting with A for AT-TACK airplanes and ending with W for WARNING airplanes. Experimental research airplanes are added on at the end. Each section contains a short introduction discussing planes of that type generally in chronological order of their appearance. The introduction is followed in each section by photos of the airplanes with captions providing basic historical information and some major specifications such as installed power, size, weight, and high speed. Photos are arranged generally in alphabetical order of the identification letters assigned by the Navy to the various airplane manufacturers. A list of assigned company letters is provided. For instance the letter F was assigned to Grumman. Within this company letter category the arrangement of photos and their captions is ordered chronologically with a few exceptions. The arrangement presents, in each section, the planes put out by each manufacturer in order, but there is a jump in time of aircraft appearance forward or backward going from one manufacturer to the next.

The designation system for Navy airplanes has changed over the years, and can sometimes be confusing. In the earliest days there was no organized number or letter system, and the model designations used by the manufacturers were sometimes used. The same was true of airplane popular names; none were assigned by the Navy until World War II, though some company assigned names were used informally. In early 1922 a Navy airplane designation system emerged. Letters were assigned for each mission (F for fighter, O for observation, P for patrol, N for trainer, and so on) and letters were also provided to each Navy airplane manufacturer. (for instance A for Fokker, later for Brewster, B for Boeing, D for Douglas, F for Grumman, etc.) A list of these letter assignments is included. So the first fighter by Grumman was an FF-1 with the second F denoting the manufacturer. A minor change would make it an FF-2. A completely new model would be an F2F-1; another brand new model an F3F-1 with its minor change an F3F-2 and so on. As will be seen some missions and their mission letters died out and new ones appeared over time along with new letters for builders as well. For a time dual missions were popular and two letters were assigned (SB for scout bomber for instance ; the first scout bomber from Vought was the SBU-1, a minor change made it an SBU-2, but a second whole new design was an SB2U-1 with the U standing for Vought in all cases.) The prefix letter X meant experimental (like XF4B-4) and V meant heavier than air airplanes not lighter than air like an airship or blimp(Z). The V was not used in airplane designations but a fighter plane was in Class VF. For a short time early on the manufacturer's letter was placed before the type letter (as a Vought UO-1 observation plane), but was quickly revised to the reverse order as shown in the designation of the next Vought observation plane design (O2U-1). There were sometimes suffix letters after the designation where an airplane was used for a special purpose (An A could mean several things, including armed, amphibian, etc.) There were many of these; they will not be covered here.

In 1962 the Department of Defense opted to change aircraft designations such that the Navy system of prior years was ended, and all current and future Navy planes were integrated into the US Air Force system. Essentially this meant dropping the Navy manufacturers letter (A Navy Douglas A3D-1 became an A-3, a Grumman A2F-1 became an A-6, and a McDonnell F4H-1 an F-4.) Modifications were delineated by a succession of letters in alphabetical order and Navy planes had different suffix letters than those of the USAF. All the nuances of this redesignation system are available in other places and will not be discussed here. The 1962 system is used today.

A particularly interesting aspect of most Navy airplanes, certainly those shipboard types, is the set of unique design and operational problems encountered as opposed to those of the Army, USAAF, and USAF, all engendered by their use in proximity to the sea. A huge factor has always been the problem of salt water corrosion on metal from sea spray. It was quickly found, for instance, that magnesium parts were very susceptable to corrosion and could not be used. Fresh water wash-downs had to be employed in any case. Space and weight limitations as well as fire hazards were always in the picture when ship based planes were being considered. In the early days right after World War I the Navy had to work out all the problems of airplane-ship compatibility, or perhaps the lack thereof. One of the first concepts of operation involved attempting to fly landplanes from Navy capital ships. They could be flown off special platforms of limited length built on the ships, and this approach was tried from battleships in the early 1920s by building wooden platforms over gun turrets for small plane takeoffs. After many experiments this approach was discarded. Ship effectiveness was compromised, and aircraft size and weight were very limited, and the wheeled airplane could not land back on the ship or in the water alongside. Landplanes were then equipped as seaplanes with floats so they could be carried aboard ships and deposited into the water and taken from it by ship's cranes. This operation was also attempted in the early days from battleships and smaller combat craft. In this concept the ship had either to move slowly through the water or stop altogether for aircraft launch and pick-up, not a very satisfactory solution in a wartime or a stormy situation. Two solutions were arrived at. One was employment of a special aircraft carrier ship with a flat deck long enough to allow airplane takeoffs and arrested landings aided by a relative wind generated by ship's speed along with ship space to hangar the assigned aircraft. The other was arranging for catapult takeoff by seaplanes from large combat ships such as battleships and cruisers with later landings in the water alongside to be hoisted aboard, both always a tricky operation. Ship catapults of sufficient power had to be designed and installed, this aspect being a story in itself. These two concepts were developed from the 1920s on through World War II. The specialized aircraft carrier is, of course, still in use, but the catapult plane concept was finished shortly after the end of World War II. Long range radar aboard ship supplanted use of catapult scout and observation seaplanes. Further, having aircraft aboard was a distinct disadvantage in several ways for big combat ships as the war had demonstrated, especially in sea battles. The aircraft were hard to handle, were always a fire hazard, and took up considerable space. The aircraft carriers on the other hand are highly developed and are now the centerpiece of naval task forces.

Carrier planes of course encounter special design problems; they need extra strong landing gear, arresting hook installations for landing, catapult hooks for those needing carrier catapulting, folding wings and sometimes tails; there are definite weight limits due to deck strength limitations, special tiedown provisions are required, and many special requirements apply. Catapult planes of the 1920s through 1940s also had special design requirements including flotation equipment for many models early on, wing folding for storage in the limited space of cruiser hangars, aircraft weight limits due to catapult power limitations, special plane trap hooks on floats, built-in hoisting harnesses, and so on. The temptation was strong to make amphibians out of seaplanes since if the amphibian also had arrestor gear it could operate not only from land or water but also from aircraft carriers and thus provide the widest possible operational flexibility. Many of the earlier seaplanes were so modified.

The progress of Naval aircraft design has been tremendous in the last 80 years. Huge advances have been made in such areas as aerodynamics, structures and materials, airframe design, powerplants, and systems and equipment. The mysteries of transonic and supersonic flight have been solved using advanced research and test facilities. New wing planform and airfoil shapes have been devised along with specially contoured area ruled fuselages. Variable geometry airframes including engine inlets have allowed optimum aircraft performance over a wide flight speed range. New means of achieving good low speed control have been found, and new structural concepts and materials have allowed very clean aircraft with excellent performance characteristics. New flight control concepts including fly by wire and by light allow implementation of optimum stability and control characteristics. The double bay wing strut and wire braced biplanes of the early years have been supplanted by sleek low drag cantilever monoplanes. Navy airplanes started with fabric covered wooden framed structure, progressed through steel or aluminum welded or bolted members with wood formers and stringers and fabric covering, and then went to all metal semi monocoque stressed skin construction. Today there is increased use of sandwich structures made up of metal or composite materials providing simplification and great stiffness.

Navy airplanes originally had heavy water cooled piston engine powerplants, then in the later 1920s started using simpler air cooled radial piston engine powerplants with less weight and vulnerability. Piston engine supercharging was developed to a high degree. After World War II the Navy started using planes with gas turbine powerplants and after a difficult initial period was able to bring airplanes equipped with these to the carrier flight deck. Today Navy tactical aircraft are turboprop, jet, or turbofan types.

In addition a revolution in subsystems has occurred in all areas, including cockpit controls and displays, hydraulic and electrical utility subsystems, and flight controls and environmental control systems. Great strides forward in aircraft avionics and weaponry have often made the aircraft simply a platform on which to carry these items. New missiles and smart bombs allow delivery of these sophisticated weapons precisely upon desired targets from stand-off distances, and long range radars mounted on Navy aircraft can provide detailed information on enemy force distributions.

The extreme cost of new aircraft types is a sobering factor and leaves open the question of the future of many types of Navy manned aircraft. There is a constant battle with the requirements of other services for development and production funds for new aircraft. The Navy has recently had a series of setbacks which can make one wonder about the future of a proud service.

NAVY ATTACK AIRPLANES

After World War II the Navy decided to combine all functions of the earlier scout bomber, torpedo bomber, and the short-lived bomber torpedo category into a simple ATTACK designation, this to provide all the offensive punch for which an aircraft carrier was designed. Further, the advent of new larger carriers, starting with the Midway class, allowed operation of bigger more powerful aircraft in a heavy attack class as well as those for light attack, the former to extend the offensive striking power of carrier task forces.

Carrier attack aircraft can be divided roughly into light and heavy attack types, and in the post World War II era a total of thirteen distinct aircraft have been designed for operation by Navy or Marine squadrons; eleven of these were produced in quantity, seven light and four heavy types.

The first two aircraft in the ATTACK category were of the single piston engine variety, the Douglas Skyraider (first designated BT2D-1, then AD-, and finally A-1 in 1962) and the Martin Mauler (first the BTM-1, then AM-1) starting in the late 1940s and going into the 1950s. Production and operation of the Mauler was limited, but length of Navy Skyraider service as the standard light attack aircraft was long. Both these planes carried bomb or torpedo loads externally. Along with a modified World War II Vought Corsair fighter having increased external store capacity and designated an AU-1 light attack aircraft, the Skyraider served in Korea and the two types delivered the majority of the carrier strike bomb loads in that conflict.

Another single piston engine light attack type employed in the late 1940s to the mid-1950s was the Grumman Guardian (initially the XTB3F-1, then AF-2). With increased emphasis on the ASW mission postwar the Guardians were formed into two-airplane hunter/search and killer variants, the first with radar and the second with torpedoes or bombs.

The first twin engined heavy attack aircraft used on carriers was the North American Savage (the AJ-) with a boost from a jet engine for dash speed and nuclear weapon carrying capability in an internal bay, this type serving well into the 1950s.

An abortive design and test effort of the early 1950s was the Douglas Skyshark (XA2D-1) single engine single seat turboprop powered light attack plane as a follow-on to the Skyraider, this development effort doomed by difficulties with the complex new XT40 twin power section turboprop engine driving dual propellers. The development failure of the same XT40 engine was a major factor in cancellation of the North American XA2J-1 twin engined heavy attack airplane of the early 1950s originally destined to be a follow-on to the Savage, but dropped in favor of the 1950s Douglas twin jet Skywarrier heavy attack aircraft.

The Navy aimed for a fast all-jet powered attack force with the mid-1950s introduction of two new carrier types, the light single jet engined Skyhawk (A4D-, later in 1962 the A-4) and the heavy twin jet engined swept wing Skywarrier (A3D-, later in 1962 the A-3). The Skywarrier really put the Navy in the big time of strategic bombing capability from new larger aircraft carriers with a nuclear store or conventional bombs in a large enclosed fuselage weapons bay and folded to be compatible with both new and older updated carriers. From the mid-1950s well into the 1960s the 600 mph Skywarrier was the Navy's big punch. These aircraft were adaptable to many other missions as well, and with the advent of Navy ballistic missile submarines the Skywarriers were modified for reconnaissance.

The small A-4 Skyhawk started as a light day attack aircraft in the middle 1950s, but served for 20 years frontline into the 1970s with continually improved capabilities including aerial refueling, all weather operation, and nuclear store delivery. Navy and Marine Skyhawks were used throughout the Vietnam War.

The Navy got a carrier based supersonic heavy attack bomber in the early 1960s in the form of a long lean 30 ton twin jet North American Vigilante (A3J-1, later in 1962 A-5) with capability of loosing an atomic weapon rearward as it flashed along at high speed. Shortly though, because of a mission change, the Vigilantes were revised for high speed reconnaissance until out of service in the 1970s. Some were utilized in the Vietnam conflict.

In 1963 another new twin engined two seat attack plane entered Navy carrier service; the Grumman Intruder (first A2F-1, then in 1962 A-6) was an A-1 Skyraider replacement. A very flexible subsonic jet with stores carried externally, the Intruder had all-weather capability and was employed extensively in Vietnam. A long-lived type, the A-6 didn't end a front line carrier career until the end of 1996.

The Navy procured the single place Corsair II light attack subsonic jet (the A-7) as a replacement for aging A-4 Skyhawks in the late 1960s, and A-7s also served in the Vietnam War. The Corsair II was a much-revised variant of the F-8 Crusader fighter and would not be replaced until the FA-18 Hornet came into service.

The AV-8 Harrier, a British basic design, was a light attack airplane for the US Marine Corps with the unique capability of V/STOL operation and special air maneuvering capabilities. Operational by the mid-1970s, the Harrier type has been continually updated.

The F/A-18 Hornet from McDonnell Douglas was designed to combine the capabilities of attack and fighter aircraft, and is discussed under the fighter category. It has served for about 17

DOUGLAS AD-1 (MFR.) A production redesignation of the XBT2D-1 from bomber-torpedo to attack category, 548 Skyraiders were ordered in April of 1945, but postwar cutbacks reduced that number to 277 including 242 plain AD-1s delivered in 1947. A big seven ton gross weight single place airplane with very good flying qualities and bombs carried externally, the design embodied many lessons learned from World War II. This Douglas El Segundo photo of May, 1946 shows an AD-1 with folded wings, Curtiss propeller, and a bomb under the fuselage.

years with the fleet. A new larger Super Hornet (F/A-18E and F) type with more capability is currently under test.

As noted the A-6 Intruder left fleet service in December of 1996, and leaves the Navy facing a big problem. In the 1980s work was under way on an A-12 radar-evading replacement type as a carrier attack plane, but in 1991 this program was cancelled because of poor progress and cost over-runs, leaving a big hole in the attack plane picture. The Navy has converted some F-14 Tomcat fighters into temporary attack planes, but the F-14s are getting old. The new Super Hornet, if procured, would probably be the attack plane of the next few years for US carriers—if it can be afforded.

DOUGLAS AD-1Q (MFR.) Thirty five of the post-war Navy order were radar countermeasure aircraft with a second crewmember aft in the spacious fuselage in charge of radar sensing and jamming equipment. AD-1 airplanes were powered by the big Wright R-3350 Cyclone of 2500 horsepower and had a high speed of about 360 mph. A window in the rear crewmember's compartment shows just aft of the folded wing in this El Segundo picture. The propeller is now an Aeroprop.

Douglas AD-1W (USN) One of the XBT2D-1 airplanes was revised to an AD-1W airborne early warning version without armament and with a radar antenna enclosed in a large radome under the fuselage belly. Two radarmen were placed at inside rear fuselage stations. This airplane was the prototype for many AEW versions to follow. Taken in January, 1948, this Navy Patuxent River photo shows how minimal the ground clearance of the radome was.

DOUGLAS AD-2 (USN) The Navy liked the AD-1, and in Fiscal 1947 ordered 156 more Skyraiders as the AD-2 with an R-3350 uprated to 2700 horsepower, structural beefup, revised landing gear doors, and additional fuel along with more wing store stations. The new models came into use in 1948. Pictured is a new AD-2 at Patuxent River equipped with a load of rockets and bombs.

DOUGLAS AD-2Q (MFR.) In addition to the AD-2 order the USN procured in Fiscal 1947 22 AEW variants of the AD-2 with a crew of two. The rear crew station just had to be somewhat claustrophobic!. This photo of the radar countermeasures aircraft was taken in August, 1948.

DOUGLAS AD-3 (USN) The Korean War showed the AD- type to be a very important ground attack aircraft. Fiscal 1948 Navy procurement included 125 of an AD-3 version with changes in the landing gear, structure, and a new cockpit canopy design. The photo shows AD-3 aircraft from USS Boxer (CV-21) over the Sea of Japan returning from an attack on a bridge west of Wonson, Korea in August, 1951.

DOUGLAS AD-3E (USN) Two AD-3 airplanes were revised to an AD-3E special electronics version for testing. This Navy photo of January, 1950 shows one with the big belly radome and an electronics pod on a wing store station. Additional vertical tail area was added on each horizontal surface to counter forward destabilizing volume of the radome.

DOUGLAS AD-3N (USN) Fiscal 1948 Navy orders included 15 AD-3N night attack versions of the Skyraider with the 2700 horsepower Wright R-3350 water injected engine, a normal gross weight of nine tons, and a high speed of about 320 mph. A post-Korean War picture of August, 1956 shows wing stations loaded with an unidentified store and the big high activity factor Aeroproducts propeller up front.

DOUGLAS AD-3Q (MFR.) Each new model of the Skyraider had its radar countermeasures variant. Twenty-three AD-3Q airplanes were procured with Fiscal 1948 funds. The picture shows the two-place -3Q variant, little different from the -2Q except for airframe revisions at the Douglas factory in May of 1949. A radar pod is carried on a port wing store station. The cockpit appears relatively small indicating the AD- was a big airplane.

DOUGLAS AD-3W (USN) One of 31 AEW versions of 1948-1949 under test at PAX River in November, 1949. The flight view shows the heavily pregnant aspect of this Skyraider version with the big radome, the external fuel tanks on wing stations, and the extra stabilizing fins on the horizontal tail. The tail hook shows this as a carrier airborne early warning aircraft.

DOUGLAS AD-4 (W. Bodie) Starting with Fiscal Year 1949 procurement authorized for 236 aircraft, later increased to 372, the AD-4 model started deliveries that year, and stayed in production until 1953 in several versions. It was used in the Korean conflict with great success. Cockpit improvements included an autopilot and advanced radar. The photo shows an AD-4 in the spring of 1950. This model was also exported to the UK and to France.

Douglas AD-4B (MFR.) Bombed up and ready for action, an AD-4B is shown in July, 1953 at the El Segundo plant near the end of the production run. Dive brake hinges for this single-seater can be seen just aft of the wing and a test instrumentation boom is fitted. The B was a special weapons version, this meaning carriage of a nuclear store under the fuselage was part of a substantial variety of weapons loads. Maximum takeoff gross weight was 12 tons; speed was 300 mph at sea level.

DOUGLAS AD-4N (Author) A civil -4N variant at a 1984 airshow with centerline drop tank, bombs, and rockets mocked up along with four 20mm cannon in the wings taxies out for a flight demonstration. This was the night attack version of the -4, 307 of which were produced in 1950, some being the NA (night version revised for day attack) and NL (winterized) subvariants. FY 1949 funds procured 29 AD-4Ns, FY 1950 36 more; later years included others.

DOUGLAS AD-4NA (USN) The photo shows two of 100 converted -4NA stripped night attack Skyraiders of Attack Squadron 822 out of NAS New Orleans as designated by the X on their tails. There were probably more store stations on the AD- airplane than most any other; count them! A lone drop tank occupies one station on these aircraft. The 1962 redesignation was A-1D.

DOUGLAS AD-4NL (USN) Flight lineup of four of 37 winterized AD-4Ns is pictured in this October, 1951 Navy photo with three of the aircraft carrying night illumination pods under the port wing. The stores under starboard wings are apparently either fuel tanks or special electronics pods.

DOUGLAS AD-4Q (MFR.) Radar countermeasures aircraft were becoming very important, and 39 of this model Skyraider were in 1949 FY Navy procurement authorizations. As this September, 1949 Douglas photo shows the Q version has only two 20mm wing cannon. An ESM pod is carried under the port wing.

DOUGLAS AD-4W (Besecker) Procurement of the carrier-based airborne early warning (AEW) version of the -4 Skyraider started with authorization for 52 aircraft in Fiscal 1949, 17 in FY 1950, and the remainder of the over 150 total came the next year. These three place Skyraiders were unarmed, had no dive flaps, and incorporated an extended canopy line. This view of Bureau Number 124771 shows the wing slat open.

Douglas AD-5/A-1E (Bowers). This Skyraider was a new design in part with a widened forward fuselage allowing side-by-side seating. Conversion kits provided for revision of the basic plane for different missions. It was longer by two feet with a much larger vertical tail, had a new dive brake, and new equipment. As many as eight passengers or four litters or a ton of cargo could fit into the rear fuselage. First flight of a -4 converted to a -5 configuration came in August, 1951. A total of 212 AD-5s were produced in 1952-3.

DOUGLAS AD-5N/A-1G (R.Besecker) One of 239 AD-5N aircraft shown in July, 1967 as the night attack version of the AD-5. The revised cabin enclosure with side-by-side seating is shown. Normal gross weight was now over ten tons with maximum weight over 12. The 2700 horsepower Cyclone gave a high speed of 320 mph.

DOUGLAS AD-5N/A-1G (MFR.) A Douglas photo of a new AD-5N shows the wider fuselage, larger vertical tail, and the standard four 20mm wing cannon. This -5 night attack version carries the nighttime illumination pod and the big fuel tank or pod underwing. These aircraft started down the line as standard AD-5s and were kit-converted downstream as needed.

DOUGLAS AD-5Q/EA-1F (MFR.) This company Santa Monica plant photo of July, 1957 shows the electronic countermeasures (ECM) version of the new Skyraider (also known as the SPAD) with underwing pods of electronic gear appropriate to the mission. Chaff dispensing capability was included. The various AD-5 aircraft were long-lived, lasting well into the 1960s.

DOUGLAS AD-5S (MFR.) One aircraft was made into an AD-5S version as shown in a Douglas El Segundo photo of June, 1953, this version equipped for one portion of an anti-submarine hunter-killer team. Underwing pods are similar to those of the AD-4NL version. The -5 arrangement made it relatively easy to modify the plane for various specialized missions.

DOUGLAS AD-5W/EA-1E (USN) An unarmed AEW version of the -5 Skyraider catches the hook for another "controlled crash" landing aboard the carrier it has been protecting. Compared to the early AD-1 the airplane, with a large search radome and wide fuselage, got very bulky.

DOUGLAS AD-5W/EA-1E (USN) An early warning Skyraider with a crew of four sits for a photograph. This version had no dive brake installation since space was taken up in the rear by two radar operators and their equipment. In any case dive bombing was not part of the AEW mission!

DOUGLAS AD-6/A-1H (P.Bowers) The next Skyraider was again a single seater for the ground attack mission and 716 of this type were procured with first deliveries in 1953 and lasting into 1956. The AD-6 was specially fitted for low level attack bombing. Details of wing folding are shown in the photo. The size of the external drop tank is noteable, as is the relatively small size of the man near the tail. The A-1 was a good sized aircraft.

DOUGLAS AD-6/A-1H (W. Larkins) The last truly mass-produced Skyraider (72 AD-7/A-1J final versions similar to the AD-6 with increased power were also delivered) is shown at an airshow. Noteable up front are the carburetor air scoop atop the cowl, the ever-present Aeroproducts propeller, and fuselage recesses for the jet exhaust stacks. The two 20mm starboard wing guns are shown, and the outline of the fuselage dive flaps can be seen. High speed was about 340 mph at 15000 feet.

DOUGLAS XA2D-1 (R.Besecker) A turboprop version of the Skyraider labeled the Skyshark for which Douglas obtained a September, 1947 contract for two prototype aircraft. Korean War pressure also resulted in a June, 1950 order for ten pre-production A2D-1s and later orders for 331 production versions. The 5100 horsepower Allison XT-40 dual power unit turboprop engine drove dual rotation propellers making for a complex powerplant installation which ended up marring the flight test program. Only eight of this 500 mph attack aircraft were test flown; four others manufactured were never flown. Contracts were terminated in late 1954.

DOUGLAS A3D-1/A-3A (P.Bowers) Largest (72'6"span) and heaviest (35 tons) of the Navy attack types, the A3D-1 Skywarrier was designed as a carrier-based twin jet strategic bomber; the X-model flew first in October, 1952 with Westinghouse J40 engines. Production A3D-1s had 9700 pound thrust Pratt and Whitney J57s giving a sea level top speed of 620 mph. Nuclear weapons could be carried. The first of 50 production aircraft started service in 1956. They were used primarily as trainers for the later A3D-2.

DOUGLAS A3D-1P (MFR.) An early photo reconnaissance version of the Skywarrier, the A3D-1P led the way for future fast reconnaissance types. The photo shows this version at the Douglas Santa Monica plant in October 1956 when Skywarriers were starting into service.

DOUGLAS A3D-1Q (MFR.) This manufacturer's photo shows a radar countermeasures version of the A3D-1. Bulges on sides of the forward fuselage and at the tail contain ESM sensor gear. This aircraft was one of the 50 production types specially modified.

DOUGLAS A3D-2/A-3B (MFR.) The definitive production model of the Skywarrier, this one at the Naval Air Test Center, Patuxent River, Md. Extent of the bomb bay can be judged by the length of the open bay door. The only defence armament was two 20mm guns at the tail. A total of 164 A-3Bs were delivered for the Navy's big carriers starting in 1957.

DOUGLAS A3D-2/A-3B (USN) Flight view of production A-3B Skywarrier shows the 36 degree sweep of the wing and podded Pratt and Whitney J57 engines rated at a maximum of 12400 pounds of thrust with water injection. High speed was 640 mph at sea level and 560 mph at 36000 feet. Overload weight at takeoff climbed to 41 tons.

DOUGLAS A3D-2P/RA-3B (USN) A photo reconnaissance version of the A-3B, deliveries of which started in August, 1959 leaves the carrier in what could be a waveoff. This model was a five place airplane with a pressurized fuselage. Tail armament was deleted. The weapons bay housed a photo-navigator and photo technician along with up to 12 different camera locations. A small unpressurized bay aft contained photoflood bombs for night work. A total of 30 RA-3Bs were procured.

DOUGLAS A3D-2Q/EA-3B (J. Weathers) An April, 1976 photo shows a seven place radar countermeasures and electronic reconnaissance variant of the Skywarrier from Squadron VAQ-308 at NAS New Orleans. The wing fold method is shown; the vertical tail also folded down for carrier stowage. First flight was in December, 1958; the A-3 had longevity. Twenty-five of this version were produced. An aerial refueling probe is on the port side.

DOUGLAS EKA-3B (AAHS) Some EA-3B aircraft were modified to include an aerial refueling tanker mission as well as ECM duties. Here an EKA-3B model of Squadron VAQ-131 from carrier JF Kennedy is shown at NAS Alameda, Ca. in the spring of 1970. Note the ECM gear pods alongside the fuselage, antenna atop the vertical tail, and modified bomb bay doors. Four electronics operators were normally carried.

DOUGLAS A3D-2T/TA-3B (AAHS) A highly modified ECM operator and bombardier trainer version of the Skywarrier from VAH-123 is shown at NAS Barber's Point. A dozen of these pressurized fuselage aircraft were used with places for a pilot, instructor, and six trainees having individual stations and equipment. This version was first flown in April of 1959.

DOUGLAS A4D-1/A-4A (P.Bowers) The A4D- Skyhawk was designed as a relatively simple light weight Navy carrier attack plane capable of delivering a nuclear store and started a very successful line of aircraft extending over decades. The first A4D-1 was flown in August of 1954 with a Wright J65 jet engine of 7700 pounds thrust. Grossing less than seven and one half tons, the A4D-1 could attain a speed of 600 mph at altitude and 675 mph at sea level. The airplane is shown with a single centerline store in contrast to most later versions.

DOUGLAS YA4D-2/YA-4B (R. Besecker) An early service test version of the A-4B Skyhawk being refueled at NATF Lakehurst, NJ. in September of 1963. The refueling system was single point. This aircraft incorporated an aerial refueling probe introduced in production on later A-4B models. The first A4D-2 was flown in March of 1956.

DOUGLAS A4D-2/A-4B (H. Andrews) A half dozen Skyhawks of Squadron VA44 aboard carrier Independence in 1960 show the relatively small size of the A4D-2. Only the near aircraft has an aerial refueling probe; early airplanes were not so equipped. Note the large wing leading edge slats extended. A total of 542 of this model Skyhawk were procured from 1956 to 1959.

DOUGLAS A4D-2N/A-4C (H. Andrews) A service trial Patuxent River A4D-2N Skyhawk is shown aboard CVA-62 with an early Phantom fighter behind. First flown in August, 1959, this version was the last incorporating the Wright J65 engine with 638 built. The 1000th Skyhawk was delivered as an A4D-2N in February of 1961. This model (it was not an A-4C until 1962) carried terrain avoidance radar and had limited all-weather capability. It also incorporated an autopilot.

DOUGLAS A4D-5/A-4E (P.Bowers) A much-revised Skyhawk, the A4D-5 first flew in July, 1961 and the next year was redesignated an A-4E. A major change was use of a Pratt and Whitney J52 engine of 8500 pounds thrust giving sea level high speeds of 685 mph clean and 575 mph with stores on five pylons. Two additional stores pylons were available under this plane. Overload gross weight with stores aboard was 24500 pounds. A total of 500 E versions were produced. The Navy conducted air strikes over Vietnam with A-4 aircraft.

DOUGLAS A-4F (J.Weathers) A Marine A-4F Skyhawk of VMA-142 is shown at NAS Jacksonville, Fl. in August of 1977 and illustrates the fact that someone in The Flying Gators could not spell! First flown in August, 1966, the A-4F shows the fuselage turtledeck hump added on Skyhawks to house more avionics. The J52 engine was uprated to 9300 pounds of thrust; 147 F versions were put out.

DOUGLAS TA-4F (R.Besecker) The initial two seat trainer version of the Skyhawk, with 238 built, had operational capability as well. A TA-4F of Squadron VA-125 is shown in the summer of 1969. As can be seen, an aerial refueling probe was installed. Use of two large drop tanks was typical.

DOUGLAS TA-4J (J.Weathers) A later two place Skyhawk trainer was the TA-4J, 293 of which were built starting in late 1969. The aircraft shown is at NAS New Orleans in April of 1977 and is from TRAWING 3. With the additional cockpit the TA-4 airplanes did not incorporate the new hump atop the fuselage.

DOUGLAS TA-4J (USN) Shown are two TA-4J aircraft used as proficiency trainers by Navy Fighter Squadron VF-126. The modified delta wing planform is well illustrated. Though Skyhawks were out of the fleet in the mid-1970s many continued operations as trainers with reserve squadrons for a long period.

DOUGLAS A-4L (J.Weathers) The A-4L was an A-4C updated in the late 1960s though still retaining Wright engines. Intended for reserve use, 100 aircraft were involved in the conversion. An A-4L of Memphis, Tenn. VA-204 is shown in November of 1976. Note the snakes on the fuselage and drop tanks along with the fuselage staining from firing the 20mm cannon.

DOUGLAS A-4M (R. Besecker) A Marine A-4M Skyhawk shown without squadron markings, one of 158 aircraft tailored for the Marine Corps using the Pratt and Whitney J52 now of 11000 pounds thrust giving a high speed of over 680 mph clean at low altitude. The A-4Ms served the Corps through the 1970s.

GRUMMAN AF-2S (R.Besecker) The AF-2S was a three place carrier based anti-submarine warfare (ASW) aircraft and was the killer part of the AF-2W/AF-2S hunter-killer team. Developed in 1948-9 from the prototype XTB3F-1 torpedo bomber, a total of 193 AF-2S models were used in the early 1950s until supplanted by the Grumman S2F- series. Powered by the Pratt and Whitney Double Wasp, the aircraft spanned 60 feet, weighed 10 to 11 tons gross, and had a high speed of about 270 mph. The belly weapons bay could house a variety of ASW weaponry such as a torpedo.

GRUMMAN AF-2W (MFR.) The other half of the carrier based hunter-killer team, the AF-2W searched for submarines using a radar antenna mounted in the big belly radome shown. The destabilizing effect of the radome was countered by the addition of small vertical tail surfaces. A total of 153 AF-2W aircraft were used by the Navy in the early 1950s.

GRUMMAN AF-2W and AF-2S (MFR.) A Grumman photo shows the ASW team of two aircraft. Interestingly the AF-2S also carries the additional vertical tail surfaces. The searchlight under the AF-2S port wing is for night illumination of a surfaced sub; a radar pod is under the starboard wing. Using two aircraft for the carrier based ASW mission was cumbersome and the problem was later solved by the Grumman S2F- Tracker.

GRUMMAN A2F-1 (F.Sullivan) The A2F-1 Intruder was the initial designation, prior to the 1962 redesignation system, of the new Grumman A-6A carrier based attack plane designed to replace the Douglas Skyraider. The early Intruder photo was taken at Quantico, Va. in November of 1962. The crew of two sat side by side.

GRUMMAN A-6A (MFR.) The Intruder shows off its load carrying capacity with thirty Snakeye bombs on multiple racks hung from five store stations. Loads up to eight tons were practical. Gross weight approximated 25 tons. The pilot was accompanied by a bombardier-navigator who operated the DIANE navigation/attack system. Power was supplied by two Pratt and Whitney J52 jet engines each of 8500 pounds thrust.

GRUMMAN EA-6A (R.Besecker) An electronics countermeasures version of the Intruder, a total of 21 built or converted from A-6As, is shown at a May,1969 airshow at MCAS Cherry Point, N.C. The large bulge atop the vertical tail contained radar sensing elements. The EA-6A was a four place aircraft including electronics systems crewmen. Radar jamming pods could be carried underwing.

GRUMMAN NA-6A (MFR.) A manufacturer's front view of an A-6A shows the large cheek-type engine air inlets and wing store stations along with a nose test boom. The N prefix designates an aircraft that has been modified such that it cannot be easily returned to its original configuration. The special test work is unknown, but probably involved new weaponry.

GRUMMAN EA-6B (R.Besecker) The EA-6B Prowler, 170 of which were produced by Grumman starting in 1971 was the definitive electronics countermeasures (ECM) version of the Intruder. A whole new fuselage enclosure for crew members, including three ECM crewmen, was involved. A large sensing antenna system was located on the vertical tail . The photo shows an EA-6B with wings folded; store stations show only external fuel tanks.

GRUMMAN EA-6B (USN) An EA-6B Prowler is shown coming aboard the carrier USS Ranger in May of 1975 in landing configuration with hook out and wing slats and flaps deployed. The white helmets of three of the four crewmen are seen. The aircraft is carrying an ECM pod and a drop tank under the starboard wing. The fixed aerial refueling probe is directly in view of the pilot. Normal gross weight was 27 tons.

GRUMMAN A-6E (J.Weathers) An advanced version of the A-6A, the A-6E Intruder had a new multi-mode radar and computer along with other updated avionics. First flight took place in February of 1970 with fleet deployment starting in late 1972. When A-6E deliveries were completed the total of all Intruder models came to 687 aircraft. The one shown, from VA-128, was photographed at NAS Pensacola (Sherman Field) in June of 1976. Two 9300 pound thrust Pratt and Whitney J52 engines provided a clean level flight high speed of 654 mph. Maximum catapult takeoff gross weight was 58600 pounds; maximum carrier landing weight 36000 pounds. The final active duty A-6 flight took place in December, 1996 by Squadron VA-75 from USS Enterprise.

NORTH AMERICAN AJ-1/A-2A (P.Bowers) The first really big Navy attack airplane, the AJ-1 Savage was designed in 1946-7 and flew first in July of 1948. A crew of three was employed along with two Pratt and Whitney Double Wasp engines and an Allison J33 4600 pound thrust jet engine in the tail to boost dash speed to 450 mph. Normal gross weight was 25 tons. About 50 AJ-1 Savages were used by the Navy in the early 1950s. The one running up was photographed at Downey, Ca. in August, 1949.

NORTH AMERICAN AJ-1/A-2A (MFR.) An AJ-1 cruising along the California coastline is caught by the company photographer. Wingtip fuel tanks were a standard feature on the Savage. Except for a substantial (up to almost 10000 pounds) of bombs that could be carried in an internal bay, including nuclear weapons, there was no armament carried.

NORTH AMERICAN AJ-2/A-2B (USN) One of 82 AJ-2 Savages at a 1950s airshow illustrates it was large for a carrier based aircraft of that time with a span of over 71 feet. The cockpit was noisy because propeller tip-to-fuselage clearance was small. The AJ-2 had a first flight in early 1953. The jet engine inlet was atop the fuselage, but many AJ-2 aircraft had that engine removed when front line service ended in the late 1950s and then were converted to aerial tankers.

NORTH AMERICAN AJ-2P/RA-2B (MFR.) Thirty of the AJ-2s were converted in 1952 to AJ-2P photo reconnaissance planes with cameras in the nose and weapons bay. This manufacturer's photo of the recon version was taken in August of 1954. The Savage was replaced by the Douglas A-3 twin jet aircraft.

NORTH AMERICAN XA2J-1 (MFR.) After World War II the Navy started to favor turboprop power, and along with the XA2D-1 light attack design initiated procurement of a new heavy attack aircraft, the XA2J-1 powered with two of the Allison T40 twin power section engines driving dual rotation propellers. First prototype flight took place in January, 1952. Size, weight, and high speed were similar to the AJ- aircraft but two tail 20mm cannon were incorporated. Failure of the T40 powerplant doomed the XA2J-1, and the Navy went on to the A3D- pure jet airplane.

NORTH AMERICAN A3J-1/A-5A (P.Bowers) Truly a lean machine, the A3J-1 Vigilante was a very fast (1350 mph at 40000 feet) carrier based twin jet strategic bomber. The photo shows the first production aircraft after two prototypes. A two place plane powered with 17000 pound thrust GE J79 engines, a normal gross weight of 27 tons, and a swept wing span of 53 feet, the A3J-1 could carry nuclear stores with rearward ejection capability from the bomb bay. The first of 50 planes was ready for operational use in mid-1961. Most A3J-1s (A-5As in 1962) were modified to RA-5C reconnaissance planes in later years.

NORTH AMERICAN RA-5C (R.Besecker) The definitive model of the Vigilante was the RA-5C reconnaissance version. The photo shows an aircraft of Squadron RVAH-3 in July of 1973. This Vigilante model carried a special bomb bay pod fitted with various photo, radar, or infra-red sensing equipment. A little heavier and slightly slower than the A-5A bomber, the RA-5C was used in Vietnam with considerable success.

NORTH AMERICAN RA-5C (USN) An excellent view of a Vigilante being readied for launch aboard the carrier USS Kitty Hawk, CV-63 in April, 1975 during a joint US, Canadian, Australian, and New Zealand naval operation. Not much of the wing folded; the vertical tail also folded down. The lower section of the reconnaissance pod can be seen below the slim fuselage. A total of 111 Vigilantes were built.

MARTIN AM-1 (P. Bowers) A post World War II redesignation of the XBTM-1 bomber-torpedo plane, the AM-1 Mauler of 1947 used the Pratt and Whitney 28 cylinder R-4360 Wasp Major engine of 3000 horsepower giving the eleven ton attack plane a high speed of about 330 mph at medium altitude. A total of 149 AM-1s were procured. This Martin aircraft was armed with four wing-mounted 20mm cannon. In 1950 the AM-1 went to the reserves and the Navy standardized on the Douglas Skyraider.

MARTIN AM-1 (MFR.) The Mauler was noted for its ability to carry heavy loads, all externally. A variety of bombs, rockets, and torpedos could be carried under wing and fuselage. As many as three torpedos could be carried, one on centerline and one under each wing. Some Maulers served awhile in the Atlantic Fleet, but first line operations were limited.

VOUGHT AU-1 (MFR.) A low altitude version of the F4U-Corsair specifically adapted to the ground attack mission, and originally an F4U-6. A total of 110 AU-1s were procured for the USMC with an initial model flying near the end of 1951. Powered by a low altitude version of the Double Wasp engine with a single stage supercharger, the heavily armored AU-1 could carry 10 rockets or two tons of bombs along with four 20mm cannon.

VOUGHT A-7A (J.Weathers) An A-7A Corsair II attack aircraft of Naval Reserve Squadron VA-303 is shown at NAS New Orleans in March of 1977 visiting from its base at NAS Alameda,Ca. Based on the F-8 Crusader fighter and to be used as an A-4 replacement, the A-7A first flew in September, 1965. Squadron service began in October, 1966 and the aircraft went into Vietnam combat in December, 1967. A total of 199 A-7As were delivered.

VOUGHT A-7B (J. Weathers) This A-7B of Squadron VA-305 was photographed at NAS New Orleans in February, 1981. The A-7B was a developed version with a Pratt and Whitney TF30 turbofan engine of 12200 pounds thrust. Without stores the Corsair II had a high speed of about 680 mph on the deck and normal gross weight was about 15 tons. Basic armament was two 20mm cannon. A varied load could be carried on racks both underwing and aside the fuselage. All 196 aircraft had been delivered by May of 1969.

VOUGHT A-7E (Author) The final production version of the Corsair II. The first 67 aircraft of the production run were still Pratt and Whitney powered and called A-7C models. Well over 500 of the A-7E type were produced using the Allison TF41 turbofan of 15000 pounds thrust. A single M61 high rate of fire cannon was used on this model and electronics systems were updated. The A-7E shown was photographed at NAS Willow Grove, Pa. in September, 1979.

VOUGHT A-7E (USN) A well used A-7E of Attack Squadron VA-72 aboard the carrier USS Kennedy shows well the fine visibility afforded by the forward cockpit location, at least to the front. A single store is carried under the starboard wing with two inboard stations empty. The A-7E saw Vietnam action starting in the spring of 1970, and served on carriers into the late 1970s and early 1980s until replaced by the FA-18 Hornet.

Hawker-Siddeley AV-8A (J.Weathers) An AV-8A Harrier V/STOL light attack aircraft of Marine Squadron VMA-231 is parked on the hardstand at NAS New Orleans in November of 1976. The Marines wanted the Harrier because of its versatility in forward area operations. Powered by a Bristol Pegasas turbofan of 21500 pounds thrust, the Harrier could operate vertically at weights somewhat less than that thrust and in STOL mode at weights up to 12 tons. The Harrier entered service in 1971. It has been continually dogged by accidents.

MCDONNELL DOUGLAS AV-8B (MFR.) A substantially improved Harrier using a Rolls Royce turbofan engine of 21500 to 23800 pounds of thrust to give greater VTOL capability along with improved structure, avionics, and armament (a 25mm high rate of fire cannon was installed), the service test YAV-8B first flew in late 1978. First flight of a production article was in November, 1981. High speed at sea level is about 675 mph. An automatic vertical landing took place in December of 1982.

NAVY BOMBING AIRPLANES

The pure bombing aircraft designation for the Navy started in the early 1930s and ended essentially with the start of World War II. Only seven distinct Navy types were included in this category during the above decade; only three could be called production aircraft. They were important in introducing the carrier based dive bomber with capability of delivering a 1000 pound bomb during a dive by aiming the plane at the target during the dive, releasing the bomb, and pulling out of the dive after release. They were also designed to be capable of pulling out of the dive without a bomb release. Two of the aircraft designated in this category were horizontal bombers, however.

Interestingly the two initial dive bombers were ordered in 1928 as XTN-1 and XT5M-1 torpedo planes but the winning Martin production design became the BM- series fixed gear biplane dive bombers, the first operational types for carriers.

In 1932 the Navy acquired a single unusual plane as a horizontal bomber prototype, the Consolidated XBY-1 high cantilever wing all metal monoplane version of the civil Fleetster light transport with an internal bomb bay in place of a passenger compartment. Its 50 foot span wing did not fold and it was thus not carrier-compatible. Perhaps the intent was use by the Marines on land. Another horizontal bomber, the NAF XBN-1, remained only a design project.

In the early 1930s the Navy sponsored another biplane dive bomber design competition between the Consolidated XB2Y-1 and the Great Lakes XBG-1 prototypes, both with two row Twin Wasp Junior engines. Both were based on a BuAer design and carried a 1000 pound bomb. The Great Lakes airplane won the production contract and BG-1s entered service in 1934 and were still being used by the Marines as late as 1940. Bell had taken over the BG-1 project from Great Lakes in 1935-36.

The Great Lakes XB2G-1 biplane 1000 pound dive bomber prototype was an abortive attempt to improve the BG-1 with new features such as a retractable landing gear and an internal bomb bay. There was no production.

In 1937-38 an important new low wing all metal cantilever monoplane dive bomber arrived on the scene. The Northrop BT-1 became a production type after some delays including a plant strike, and through modification of the final aircraft on contract the design was revised to an XBT-2, the importance of which became obvious as it was the fore-runner of the famous World War II Douglas SBD-Dauntless dive bomber series.

The final aircraft in the Navy bomber series was the Douglas BD- type, a version of the Army A-20 Havoc light horizontal attack bomber. Of the few procured by the Navy none was ever used in a bombing role but rather employed in utility duties.

The Navy bomber series of aircraft were important in the early development of the dive bombing technique utilized later in World War II carrier based aircraft.

DOUGLAS BD-2 (USN) With the Army ordering the A-20 Havoc type as a light bomber the Navy indicated interest in the twin engined aircraft, and evaluated one A-20A as a BD-1 in December, 1940. A further eight A-20B Havocs were procured in the spring of 1942 and designated BD-2s. They were employed as utility aircraft by the USMC since the planes were not suited for Navy combat missions. Two Wright R-2600 engines gave the BD-2 a high speed of about 350 mph.

BELL XBG-1 (H.Thorell) The winner of an early 1931 competition for a dive bomber able to carry a 1000 pound bomb load, the XBG-1 was a Great Lakes airplane contracted for in June, 1932 and conforming to a Navy Bureau of Aeronautics design. The 1933 prototype shown was a near three ton gross weight two place conventional open cockpit biplane with a Pratt and Whitney R-1535 engine giving a high speed of about 180 mph. Armament complement was one fixed and one flexible .30 caliber machine gun. The photo shows the XBG-1 after enclosing hatches were fitted over the cockpits.

Bell BG-1 (Swisher) The Great Lakes Corporation went out of business in 1936, and the new Bell Aircraft Corporation took over BG-1 production after three Navy contracts from November, 1933 through February, 1935 resulted in orders for 60 aircraft. Shown is a BG-1 of Navy Squadron VB-4 with the diving tiger insignia on the fuselage. Production aircraft had cockpit enclosures.

BELL BG-1 (H.Thorell) Another BG-1 of VB-4 is shown with wing racks for light bombs and a swinging crutch pivoted just aft of the engine cowling to allow the 1000 pound bomb beneath the fuselage to clear the propeller in a dive. The BG-1 lasted until 1940 with the Marines and was finally replaced with monoplane Dauntlesses.

GREAT LAKES XB2G-1 (USN) In 1934 the Great Lakes Corporation was requested to produce an updated BG-1 type with the same Pratt and Whitney R-1535 engine of 750 horsepower and the same wings with a new fuselage deepened to enclose a retractable landing gear and an enclosed bomb bay. With performance hardly greater than the BG-1 the new aircraft was not put into production. Shown is a March, 1936 Navy photo of the XB2G-1 not long before Great Lakes Corporation failed.

GREAT LAKES XB2G-1 ((USN) Navy flight photo of the XB2G-1 in March of 1936 shows the portly fuselage resulting from retracting the gear and enclosing the big bomb. The single prototype ended up serving as a Navy Command plane. It was overshadowed by development of such monoplanes as the Northrop XBT-1 of 1935 carrying the same weapons.

MARTIN BM-1 (R.Besecker) A redesignation of Martin's XT5M-1, the BM-1 was designed specifically as a 1000 pound dive bomber and 12 were ordered in April of 1931 with acceptance in September. Four others were ordered later. The aircraft had all metal fuselages with a swinging crutch for the bomb. The planes served dive bombing squadrons from 1932 to 1937; they were the first practical Navy dive bombers. The plane shown has an external fuel tank.

MARTIN BM-2 (H. Thorell) Slight modifications from the BM-1 defined the BM-2 dive bomber using the same engine, a 625 horsepower Pratt and Whitney Hornet which gave a high speed of about 145 mph. Gross weight was 6200 pounds. The armament complement was one each fixed and flexible .30 caliber gun. Sixteen BM-2s were delivered starting in August, 1932.

MARTIN BM-2 (J. Davis) Flight view of a BM-2 dive bomber shows the rear cockpit .30 caliber gun. The BM- aircraft were the first planes built in the new Martin plant near Baltimore Md. after a move from Cleveland. After 1937 the BM-2s were used as utility planes. One was still at the Naval Aircraft Factory in Philadelphia in 1940.

NORTHROP XBT-1 (USN) A Navy photo taken in April of 1936, about four months after delivery, shows the low wing all metal two place monoplane configuration of Northrop's new dive bomber with retractable landing gear folding into substantial bulges underwing. One of the earliest monoplanes slated for production, the XBT-1 had a Pratt and Whitney R-1535 engine of 700 horsepower, gross weight between 6000 and 7000 pounds, and a high speed of just over 200 mph carrying its design 1000 pound bomb under the fuselage.

NORTHROP BT-1 (MFR.) A production BT-1, probably the first of 54 ordered in late 1936. The date of the photo is September, 1937. A strike at Northrop held up delivery of the BT-1 airplanes until 1938. As can be seen the landing gear did not retract flush into the wing. Armament was one each .50 caliber and .30 caliber machine gun. A new feature was wing flaps to be used for dive bombing split into upper and lower sections with perforations to prevent vibration.

NORTHROP BT-1 (H.Thorell) A BT-1 dive bomber from Navy Squadron VB-5 runs up prior to a takeoff. The rear gunner appears to be missing. Tail hook, radio aerial, and open cowl flaps are in evidence. Elevators are up to catch slipstream and keep the tail from jumping. The production aircraft had an 835 horsepower Pratt and Whitney R-1535 engine. One BT-1 was reconfigured for testing the then-new tricycle landing gear.

NORTHROP XBT-2 (MFR.) One of the Northrop production BT-1 airplanes was held back for major modifications into a cleaner more powerful dive bomber; the basic changes involved use of a Wright R-1820 Cyclone engine of 1000 horsepower and a new flush retracting landing gear. A new tail shape and other changes made the XBT-2 finally evolve into the first of the famous Douglas SBD- Dauntless dive bombers. High speed with a 1000 pound bomb was about 250 mph.

CONSOLIDATED XBY-1 (MFR.) The XBY-1, Consolidated Model 18, was a Navy level bomber revision of a commercial high wing monoplane Fleetster. Unique in being of all metal monococque construction with a cantilever wing containing an integral fuel tank, the XBY-1 had a 50 foot span non-folding wing and couldn't dive bomb. Its Wright Cyclone engine allowed a top speed of 170 mph and gross weight was 6600 pounds. Only one aircraft was built.

CONSOLIDATED XB2Y-1 (MFR.) A manufacturer's photo of June, 1933 shows the XB2Y-1 with a 1000 pound bomb in the crutch designed to swing it clear of the propeller upon dive release (*center, right*). The last design by B. Douglas Thomas, it was considered too heavy and expensive with a big center fuselage part hogged out of a steel block. Powered by a 700 horsepower Pratt and Whitney R-1535 radial the XB2Y-1 had a maximum speed of 180 mph at 9000 feet. It remained a single prototype.

NAVY BOMBER FIGHTER AIRPLANES

For a short period in the mid-1930s the Navy defined a bomber fighter category of aircraft; these were basically fighter types for a primary light bombing role, and were used in carrier bombing squadrons.

In 1934 the single seat biplane Curtiss F11C-2 airplanes, with performance little better than the late Boeing F4B- fighter types, were re-categorized as BFC-2 bomber fighters for light bombing squadrons with the biplane F4B-4s and Grumman F2F-1 biplanes serving the fighter squadrons.

Shortly after the advent of the BFC-2s the final production plane of that series was modified to a retractable landing gear version, first as the XF11C-3 but shortly redesignated as a BF2C-1 light bomber fighter. The aircraft served briefly in carrier bombing squadrons but were discarded because of technical problems.

The experimental Boeing XF6B-1 biplane fighter prototype meant to be an F4B-4 follow-on and a much revised aircraft with a Twin Wasp engine was revised briefly on paper to an XBFB-1 bomber fighter but remained a single aircraft.

CURTISS XBFC-1 (NACA) This light bomber-fighter airplane by Curtiss was a late redesignation of the experimental XF11C-1 fighter of 1932 and used a 600 horsepower Wright two row R-1510 radial engine. This engine was not successful and the airplane remained a prototype. Redesignated as a 500 pound bomb XBFC-1 dive bomber in March of 1934, it was turned over to NACA for engine cooling test work on special cowlings.

CURTISS BFC-2 (P.Bowers) In 1934 the Navy production F11C-2 was redesignated as a light fighter-bomber, and a revision was made in the aft turtledecking along with a partial enclosure of the cockpit. The photos (*top, right*) show a BFC-2 of the Navy's VB-3 High Hat Squadron. These aircraft served until 1938.

CURTISS BFC-2 (J.Weathers) Another High Hat Squadron BFC-2 is shown in the photo (*below*); these aircraft were the last Navy fixed gear Hawks, sometimes called Goshawks. A 700 horsepower Wright nine cylinder R-1820 Cyclone gave this big-wheeled panted gear light bomber a high speed of about 200 mph, depending on load. Light bombs could be carried on wing racks; the 500 pounder was carried under the fuselage.

CURTISS BF2C-1 (MFR.) A September, 1934 photo of the BF2C-1 at the Curtiss plant shows the revision of the basic Goshawk fuselage to adapt to a retractable landing gear. The fifth production F11C-2 was modified to a retractable gear XF11C-3 which was the prototype for the BF2C-1 production aircraft. A bomb or external fuel tank could be nestled in the recess between the housings for the retracted wheels.

CURTISS BF2C-1 (MFR.) A BF2C-1 light bomber-fighter runs up its Wright R-1820 Cyclone engine at New York's Floyd Bennett Field in 1935. Gear wheel pockets and mechanisms are well shown. Unfortunately Curtiss's new metal framed wing design was sympathetic to Cyclone engine vibrations at desired operating engine RPMs. In early 1936 the BF2C-1 had to leave Navy service.

CURTISS BF2C-1 (P. Bowers) Another BF2C-1 from carrier Ranger's VB-5 light bombing squadron with external tank aboard. Shortly in 1935 these aircraft were withdrawn from service; they were to be the last Curtiss production fighter types for the Navy. Export versions of the BF2C-1s as Hawk IIIs used the earlier wooden framed wings of the BFC-2s.

NAVY BOMBER TORPEDO AIRPLANES

Starting during World War II the concept of a single place dive bombing and torpedo aircraft was advanced for carrier use by Navy BuAer. Such aircraft would replace two or three place scout/dive bombers like the SB2C- and the torpedo bombers like the TBF- and TBM- with functions combined into one aircraft, a VBT type. The idea was that experience showed both dive bombing and torpedo attacks needed fighter cover in any case, and removal of a second and maybe a third crewman along with their defensive armament would result in a smaller lighter aircraft with improved performance to do the same weapons load carrying job.

Two early aircraft prototypes in this category were the Curtiss XBTC-1 and XBTC-2 and the Douglas BTD-1. The Curtiss program was delayed by various difficulties; the plane was single place with a Pratt and Whitney Wasp Major engine driving dual rotation propellers, carried a load externally, but did not get past the prototype stage. The Douglas BTD-1 was a single place rework of the earlier XSB2D-1 design dispensing with the power turrets and aft crewman. It also featured a tricycle landing gear and used the Wright Duplex Cyclone engine. A few were produced but none were employed operationally. Performance was still less than desired and an XBTD-2 version was planned with Westinghouse jet engines in the aft fuselage but nothing came of the project.

Later contracts were signed for four different bomber torpedo types with a range of different powerplants. The competitors were the Curtiss XBT2C-1, a major single place modification of the SB2C- Helldiver design retaining an internal weapons bay, the Douglas XBT2D-1, a new design using the Wright R-3350 Duplex Cyclone and carrying loads externally, the smallest and lightest competitor, the Kaiser Fleetwings XBTK-1 with a Pratt and Whitney Double Wasp engine, and the largest of the group, the Martin XBTM-1 Mauler with a Wasp Major engine. All the aircraft had bubble canopies over a single pilot and special attention was paid to engine installation design of the large high powered radial piston engines in all these VBT aircraft. Though Martin received a small order for a production AM-1 Mauler version of the XBTM-1, the big winner was Douglas with an initial 25 BT2D-1s ordered in 1944 and appearing in 1945 and later very large orders coming post-war for the follow-on AD- Skyraider attack series.

With the advent of the post-war attack (VA) category the short-lived VBT designations ceased to exist. The important result was the Skyraider which lasted at least through the 1950s and into the 1960s.

CURTISS XBTC-2 (MFR.) One of four Navy candidates for the new role of a bomber-torpedo aircraft, the Curtiss XBTC-2 was initially contracted for in mid-World War II as an XBTC-1 and evolved with minor changes as two XBTC-2 prototypes eventually delivered in 1946. A large single seat aircraft of 50 foot span and nine to ten ton gross weight, this monoplane used a Pratt and Whitney R-4360 Wasp Major engine of 3000 horsepower driving a dual rotation propeller which together produced a high speed of over 375 mph. All bomb or torpedo ordnance was carried externally.

CURTISS XBTC-2 (MFR.) Chastened Curtiss employees gather near the scene of an accident involving an XBTC-2 prototype (*right*). Three of the four wing mounted 20mm cannon muzzles can be seen as well as the mangled blades of the dual rotation propeller. No dual propeller aircraft ever got accepted for production by the Navy; the XBTC-2 was no exception. Wing leading edge slats are deployed.

CURTISS XBT2C-1 (MFR.) Curtiss was permitted to try again in the bomber-torpedo category with an order for ten airplanes (nine were delivered) in early 1945. They appeared early the next year. Again a single place monoplane with bomb and torpedo loads carried externally, this aircraft was powered by a Wright R-3350 Double Cyclone of 2500 horsepower and a single rotation propeller, giving a top speed of 350 mph clean.

CURTISS XBT2C-1 (MFR.) Flight views of an XBT2C-1 by the company photographer shows an APS-4 radar pod under the starboard wing. The aircraft was armed with two 20mm cannon. As with the Douglas XBT2D-1 aircraft the XBT2C-1 could accomodate another crewmember aft of the pilot inside the rear fuselage. The Douglas airplane was favored by the Navy and Curtiss got no production contract.

DOUGLAS BTD-1 (MFR.) The BTD-1 Destroyer of 1944-5 was a simplified single place offshoot of the earlier two seat XSB2D-1, itself an intended replacement for the famous SBD- Dauntless dive bomber series. The photo shows a Destroyer at the Douglas El Segundo plant in early December of 1943, actually before first flight. A gulled wing and a tricycle landing gear were outstanding features. Bombs or a torpedo could be carried within an internal bay. Two 20mm cannon were in the wings.

DOUGLAS BTD-1 (USN) Initially hundreds of BTD-1 aircraft were ordered in August of 1943. The photo shows a Destroyer during a test flight near Patuxent River NATC in Maryland in early September, 1944. Orders were cut back later, and by war's end only 28 were produced. Powered by a Wright R-3350 engine of 2300 horsepower and weighing over nine tons, the BTD-1 had a high speed of about 340 mph. None were ever used in squadron service.

DOUGLAS XBT2D-1 (MFR.) The reason the BTD-1 was not produced was the XBT2D-1, a simpler, lighter more straightforward single place design with an uprated R-3350 engine of 2500 horsepower and carriage of all stores outside under the aircraft. Armament again was two 20mm cannon in the wings. Twenty-five were ordered in mid-1944 and a first flight took place early in 1945. Initial testing was very successful.

DOUGLAS XBT2D-1 (MFR.) An XBT2D-1, one of the initial 25 airplanes, is shown at the Douglas factory in June of 1948 and was used for special armament tests. The aircraft type was soon to turn into the famous AD-1 Skyraider. In clean configuration the XBT2D-1 could top 370 mph at a medium altitude. Night attack and radar countermeasures versions were also in the works.

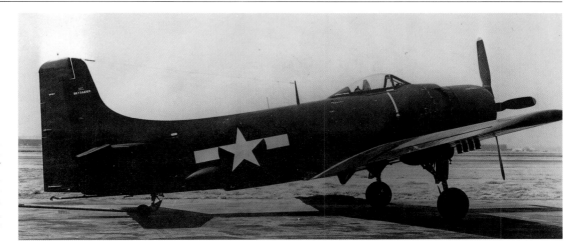

DOUGLAS XBT2D-1P (MFR.) A Douglas photo of August, 1946 illustrates the single photo version of the XBT2D-1. The aircraft was a well-liked machine with nicely harmonized flight controls and was shortly to take on varying mission roles, but no more specialized photo versions were built.

FLEETWINGS XBTK-1 (MFR.) Another competitor for a new Navy combination bomber-torpedo plane was the XBTK-1 using a Pratt and Whitney R-2800 Double Wasp engine of 2100 horsepower. Fleetwings was a division in Pennsylvania of Kaiser Cargo Inc. Weighing about six and one half tons gross with a high speed of about 370 mph, the XBTK-1 is shown in a company flight photo of July, 1945.

FLEETWINGS XBTK-1 (MFR.) A runup photo of the Fleetwings competitor shows a drop tank under the fuselage and rocket standoffs underwing. A contract for five test aircraft was let in early 1944 and first flight took place a year later. No production resulted.

MARTIN XBTM-1 (AAHS) One of two experimental models of the later Martin AM-1 Mauler and first flown in August of 1944, the XBTM-1 is shown in the markings of NATC Patuxent River, Md. carrying a full load of bombs and rockets. With a Pratt and Whitney 3000 horsepower 28 cylinder Wasp Major engine the Martin bomber-torpedo airplane was a heavy lifter. Wingspan was 50 feet; high speed was about 360 mph.

MARTIN XBTM-1 (MFR.) The XBTM-1 is shown flying in clean configuration. The long chord cowling reflects use of the big four row Wasp Major " corncob" engine. Hundreds of BTM-s were ordered during wartime in 1944, but these contracts were cut back post-war to 149 aircraft redesignated as the AM-1 Mauler.

NAVY FIGHTER AIRPLANES

US Navy fighter aircraft designs started during the last years of World War I and have continued through to the present time evolving from slow wooden piston engine biplanes to sleek all metal high Mach number missile-equipped turbine powered monoplanes.

A successful early fighter was the two place Curtiss 18-T triplane Wasp of 1918 with an excellent high speed for its time of over 160 mph. Another early design was the two place Curtiss HA Dunkirk Fighter, so called because it was designed to protect American patrol planes from German seaplane fighters. The HA was not a success. Neither of these designs was put into production, but the 18-T became a racer.

Directly after World War I the Navy experimented on fighter operations with takeoffs from platforms built on major combat ships but no facility for landing back aboard using foreign fighter aircraft designs from the war. These experiments did not lead to practical operations.

A few Army Thomas Morse MB-3 Army fighters were used, but the first types really employed as fighters were single seat versions of Vought VE-7 trainers; these were used in early aircraft carrier experiments with the USS Langley.

The initial Navy designed fighter was the Curtiss/NAF TS-1 of 1922, a convertable landplane or seaplane light single seat biplane fighter with a Lawrence radial engine for the carrier Langley. The Navy experimented on a 1924 Curtiss Hall F4C-1 version of this using basic metal construction under fabric covering as well as on a Dornier metal plane design from Germany, but did not pursue this technical advance immediately.

Two 1925 fighter designs of then conventional steel tube fuselage and wooden wings construction were the Boeing FB-1 and the Curtiss F6C-1 Hawk, both using the new Curtiss D-12 liquid cooled engine of racing plane fame. Both designs were developed through several versions in the 1920s and equipped Navy/Marine fighter and light bomber units.

In 1927 and 1928 new carrier or battleship fighter designs evolved using a significant technology advance in powerplants; the Boeing F2B-1 and F3B-1 and the Curtiss F6C-4 and F7C-1 aircraft employed the new Pratt and Whitney Wasp radial air cooled engine, this presaging Navy use of this type of engine almost exclusively for many years. Compared to liquid cooled installations the Wasp type was lighter, simpler, and less vulnerable to gunfire.

Other fighter types tested in the 1926-27 time period were the Wasp powered Eberhardt FG-1 Comanche and the Wright F3W-1 Apache, both convertable to single float seaplanes, the latter famous for setting altitude records, and the FU-1 battleship fighter converted to single place from the Vought UO- two seat observa-

tion aircraft. The latter plane was the only type put into production of these four and was Wright Whirlwind powered.

The new Wasp engine was also installed in two other 1928-29 competing two seat defensive fighters, the Vought XF2U-1 and the Curtiss XF8C-2 Helldiver. The Vought was not produced but the F8C-4/-5 Helldiver went into Navy/Marine operational squadrons. Some F8C- types were actually Falcon observation planes; in any case the two seat Helldivers were shortly all relegated to observation duties since they were found inadequate in performance compared to single seat fighters.

In 1929-30 two other new Wasp powered single seat fighter designs were tested, the Hall XFH-1 and Berliner Joyce XFJ-1/-2. Except for wing and tail covering these planes were of metal construction. The Hall design used internal flotation compartments and could detach its landing gear for emergency water landings. Neither of these designs were accepted for service use.

Boeing produced a new Wasp powered fighter prototype for carrier use in 1929; this type was to became a standard Navy service fighter well into the 1930s with follow on variants incorporating new technical features during that period. It was the F4B- series, the naval counterpart of the Army's P-12 pursuit planes.

In 1932 another two light fighters designed for aircraft carrier use appeared, both powered by Wright Whirlwind air cooled radial engines; the American Fokker XFA-1 and the Curtiss XF9C-1 both had all metal fuselages and upper wings mounted directly to the fuselage. The XFA-1 remained a prototype, but the Curtiss plane was modified into the F9C-2 Sparrowhawk and operated from rigid airships Akron and Macon until their demise.

The initial monoplane designed for Navy carrier use was the high wing XF5B-1 from Boeing with a Wasp engine. Though not put into production it had an all metal fuselage used in later F4B-fighters.

Other early 1930s biplane fighter projects for the Navy using radial engines were the Curtiss XF11C-1 and F11C-2 Hawk type single seaters with the latter put in limited production and later becoming the BFC-2, the Boeing XF6B-1 and XF7B-1 single place types, the former as a biplane modification of the F4B-, and the latter an all metal low wing monoplane of modern configuration tested at length but rejected by the Navy, and two Curtiss high wing monoplanes with retractable landing gear and radial engines, the XF12C-1 two seater and the XF13C- single place fighter were also tested, the latter capable of being reconfigured into a biplane. There was no production of any of these planes, with the Boeing F4B-types remaining as the Navy's in-service fighter. The same was true of a slick looking Berliner Joyce XF3J-1.

Other two seater biplane fighters tested in the 1933-34 time period were the Douglas XFD-1, the Berliner Joyce XF2J-1, the Grumman FF-1, and the Vought XF3U-1. The first two remained as prototype airplanes only, and the Vought fighter after a long development was turned into a production SBU-1 scout bomber. The single fighter success was the Grumman radial engined FF-1 "FiFi", Grumman's first airplane with its stubby fuselage, enclosed cockpits, and retractable landing gear. The FF-1 became an operational fighter and also spawned the SF-1 scout series.

Northrop tried to get a Navy production contract in 1934 with their XFT-1 and -2 all metal low wing monoplane with fixed landing gear. Not as maneuverable as the biplanes and with flying quality problems, the XFT- airplane did not win Navy favor.

The new middle-1930s production Navy fighters to replace the Boeing F4B- series were the maneuverable Grumman F2F-1 and F3F- series, single seat retractable landing gear biplanes with speeds in the higher 250s mph range. These were the pre-World War II Navy fighter types with fully cowled high power single or twin row radial engines. In 1939-40 with World War II in near view the Navy contracted for two new monoplane fighters in the then-modern mold, the Brewster F2A- and Grumman F4F-. Both finally went into production, the Brewster first in a modest way, and more extensively the Grumman, the latter turning into the famous wartime Wildcat as both F4F- and FM- types. A late 1930s Seversky version of the Army P-35 monoplane , the NF-1 Navy fighter prototype was also tested but rejected for poor performance.

Two prototype airplane fighter candidates , the Grumman twin engined XF5F-1 Skyrocket and the Bell XFL-1 Airabonita were tested in the 1940-41 period. The Bell was a naval P-39 version with tail down landing gear; neither was put into production.

The plane the Navy wanted for its wartime shipboard fighter was the powerful new Vought F4U-1 Corsair where the new Pratt and Whitney Double Wasp engine provided a 400 mph high speed. Unfortunately carrier compatibility problems kept this fighter as a Marine land-based aircraft until late wartime. It was also produced in volume as the Brewster F3A-1 and the Goodyear FG-1, and later in 1945 a few XF2G-1s were produced with 3000 horsepower radial Wasp Major engines.

The Navy fighter of World War II starting in 1943 and replacing the F4F-4 Wildcat was the F6F- Hellcat carrier monoplane from Grumman, a particularly successful type though gradually giving way near the end of the war to the later Corsair models.

Wartime or just postwar prototype Navy fighters that came to nought were the Curtiss XF14C-1 of 1944, and later the XF15C-1 composite piston and jet-powered type of 1945, and the big Boeing XF8B-1 meant for the larger carriers coming into service. Another composite powered type, the Ryan FR-1 Fireball, got into limited production just at war's end, but was never really successful, nor was its composite turboprop and jet powered follow-on the prototype XF2R-1 Dark Shark.

An unusual Navy prototype fighter of late wartime was the Vought XF5U-1 "Pancake" with a predicted very wide speed range. This strange machine had two piston engines remotely shaft-connected to large propellers. It never flew.

The final Navy production piston engined fighters were both Grummans, the F7F- Tigercat and the F8F- Bearcat, the latter a small high performance short range type, and the former a sleek twin engined single or two seat machine used for several missions including night fighting and photographic reconnaissance. These planes would be edged out by the jets in the late 1940s and early 1950s.

The Navy entered into the jet fighter arena with the XFD-1 Phantom from McDonnell (later changed to FH-1) and Vought XF6U-1 Pirate in 1946 just after the war. These were followed shortly by the North American FJ-1 Fury, all single place types with straight wings. One problem was that early jet engine thrust was not sufficient for carrier takeoffs without a catapult. A small production order was given for Phantoms to indoctrinate Navy fliers, and the same for the Fury, but the Vought Pirate though it pioneered new structural materials and afterburners was not a success.

More advanced straight wing jet fighters followed in the 1949-51 time period including the single engine Grumman F9F-2 Panther and the twin engine McDonnell F2H-1 (originally XF2D-1) Banshee single seaters, and the Douglas F3D-1 Skyknight twin engine two place nightfighter. All these became production aircraft and were utilized in the Korean War, the first two extensively and the Skyknight less so. All had subsequent variants, the Panther up to a -5 version and the Banshee to a -4.

The Navy sorely needed a high performance swept wing or delta wing fighter as shown by the appearance of the Russian swept wing MIG-15 in the Korean War. Several interesting approaches were taken; one of the first being sweeping the wing of the F9F-Panther and obtaining the F9F-6 to -8 production Cougars, another procuring a navalized version of the USAF F-86 Sabre as the later FJ-2 to -4 Fury series. The FJ- Furys were produced in quantity at North American, Columbus, Oh.

In 1954 the Navy projected a new concept in fighters, turboprop single seat VTOL tail-sitters with dual rotation propellers driven by the snake-bitten XT40 engine. The two aircraft were the Lockheed XFV-1 and the Convair XFY-1 Pogos. A combination of a faulty concept and an underdeveloped engine doomed the projects. For one thing the pilots could not easily back down vertically to land.

Another unusual fighter project of the early 1950s, a very prolific time, was the water-based Convair XF2Y-1 hydroski Sea Dart tested in various configurations, but finally dropped from development.

To further illustrate the prolific 1950s in terms of fighter design was the 1954 single engine single seat Douglas F4D-1 Skyray, or "Ford", a fast climbing modified delta wing interceptor put into limited production after a long development period and two years later a follow-on XF5D-1 missile armed all weather fighter, the development of which was cancelled.

Grumman in 1952 designed a big swing-wing XF10F-1 with a number of advanced features, but the result was one of only two Grumman designs never to be produced. Shortly thereafter they put out the swept wing single place F11F-1 Tiger, originally an XF9F-9. The Tiger was in moderate production but had a short service life. It was known principally as a Blue Angels aircraft.

Two mid-1950s swept wing jet aircraft had a mixed history. The 1955 twin tail Vought F7U- Cutlass suffered through a protracted development period with a -3 version seeing only limited production. It was not a success. Shortly afterwards McDonnell had a fiasco with their F3H-1 Demon development principally because of a J40 jet engine problem. The F3H-2 Demon with another engine finally developed into an operational all-weather fighter armed with missiles for the 1960s.

After two failures of Navy fighter types Vought really came through in the late 1950s and early 1960s with a very successful supersonic swept wing jet fighter, the single seat single engine missile armed Crusader (first an F8U-1, later in 1962 an F-8A). An important feature was the variable incidence wing. Late models featured limited all weather capability and guns were aboard as well as rockets. They were used successfully in Vietnam.

Competitive designs for a 1960s Mach 2 plus Navy fighter both incorporated wing boundary layer control systems to reduce carrier landing speeds were the Vought F8U-3 Crusader III and the McDonnell Phantom II (F4H-1, then F-4A). The twin engined two seat Phantom won the competition and was produced in great quantity, becoming the first line Navy missile fighter for many years, and was so successful it was also adopted by the USAF.

In the mid-1960s the Department of Defense under Robert McNamara decided an essentially common aircraft could serve both the USAF and the Navy as a swing wing turbofan fighter plane, and the result for the Navy was an F-111B variant of the Air Force F-111A. The F-111B was, among other things overweight and never became a production item.

In 1972 a new advanced fighter appeared to replace the Phantom, the variable wing sweep Phoenix missile equipped Grumman F-14A Tomcat with the ability to handle several targets at a time from a distance. This aircraft and its later variants is still the standard Navy fighter on todays carriers and together with the 1980s McDonnell Douglas FA-18A Hornet light attack fighter comprises the latest equipment the Navy has. The Hornet has been aboard carriers now for about 17 years and the Navy's hopes for the future are pinned on a newer bigger faster more powerful Super Hornet with more fuel and weapons. The Super Hornet was rolled out in September of 1995 and is now under test. The 1970s vintage F-14 Tomcats are nearing the end of their service lives. The future is certainly questionable for the US Navy fighter.

CURTISS HA (MFR.) Tested both as the landplane version shown in January, 1919 and a single main float seaplane, the two place four gun HA was developed to counter German seaplane fighter attacks on US patrol planes around Dunkirk in World War I, and named the Dunkirk Fighter. The effort was a failure. Three aircraft were built with 380 horsepower Liberty engines; speed of the two ton gross weight airplane was about 127 mph.

CURTISS 18-T (MFR.) Another early fighter, the Curtiss-Kirkham 18-T triplane fighter of mid-1918 was tested successfully and achieved a high speed of 162 mph using a 350 horsepower Kirkham-designed engine. The Navy used the Curtiss-Kirkham as a seaplane racer post-war. This airplane was the first and only Navy triplane.

FOKKER XFA-1 (USN) One of two small light fighters based on a Navy Bureau of Aeronautics design, the XFA-1 prototype, along with the Curtiss XF9C-1, was ordered in June, 1930. Powered by a Pratt and Whitney engine inside a ring cowl, the XFA-1 was delivered in March, 1932, the date of the photo. It was characterized by an all metal fuselage and an upper wing tied directly to the fuselage. Gross weight was only 2500 pounds, wingspan 25 feet six inches, and high speed 170 mph. Only one prototype was built.

BREWSTER XF2A-1 (USN) Shown in flight near NAS Anacostia in Washington, DC, the XF2A-1 shows up as a chubby fuselage mid-wing monoplane with a Wright Cyclone engine of 950 horsepower up front. Initiated as a project in 1935 along with the Grumman XF4F-2, the XF2A-1 started tests in late 1937 resulting in an order for 54 F2A-1s in June, 1938. Only a few served in the US Navy, however, since most of the order went to Finland.

BREWSTER, XF2A-2 (USN) An excellent Navy flight photo shows the XF2A-1 prototype modified into an XF2A-2 using an uprated Cyclone engine and increased vertical tail area. The photo shows how landing gear wheels neatly pocketed into the fuselage belly. The landing gear was the source of many later troubles though. The 1200 horsepower Cyclone powered the 5500 pound fighter to a 325 mph top speed.

BREWSTER F2A-2 (Curtiss Propeller Div.) One of 43 production F2A-2 fighters of late 1940 at the Curtiss Wright airport in Caldwell, NJ; the plane had a Curtiss Electric propeller and a 1200 horsepower Wright nine cylinder Cyclone engine. Details of the landing gear can be seen. The upper cowl air inlet is for the carburetor and the lower one for the oil cooler.

BREWSTER F2A-3 (USN) A Brewster fighter used as a trainer cruises over Florida in 1942. Weight gains turned the Brewster from an agile aircraft early on into a later sluggish type. Four .50 caliber guns were carried, two in the fuselage forward and two in the wings. A total of 108 F2A-3s were delivered to the US Navy in 1941. Those in the Pacific were later mauled by the Japanese.

BREWSTER F3A-1 (USMC) A pair of Corsairs manufactured by Brewster, and equivalent to Vought F4U-1As, cruise near their base, MCAS El Toro, Cal. shortly after war's end in January, 1946. Brewster put out 735 Corsairs starting in mid-1943 from a new Johnsville, Pa. plant. Some went to the British.

BOEING FB-1 (MFR.) One of ten fighter aircraft provided the Marine Corps after being built by Boeing in December, 1925, the FB-1 was very like the Army's PW-9. A feature was use of a welded steel tube fuselage frame, an advance over the then-conventional wood fuselage. The FB-1s were used by the USMC in China. A 400 horsepower Curtiss D-12 engine powered the 2950 pound fighter to a high speed of 165 mph. The FB-1s were not equipped to operate off an aircraft carrier.

BOEING FB-3 (MFR.) After the Navy procured two FB-2s like the FB-1s except adapted for carrier operations, Boeing put out three FB-3 fighters able to operate using either wheels or twin floats, a Navy requirement of that time. The planes used 510 horsepower Packard engines. The Boeing photo shows a float-equipped FB-3 in April, 1926.

BOEING FB-4 (MFR.) This version of the Boeing fighter was limited to a single airframe powered with an experimental Wright P-1 radial engine of 440 horsepower which did not prove out. It did serve to initiate the Navy trend towards the air-cooled radial engine type instead of water-cooled powerplants. The landplane version of the FB-4 is shown in January, 1926.

BOEING FB-5 (MFR.) A Boeing photo of September, 1926 shows the "production" version of the Navy fighter, one of 27 ordered and later delivered to the first aircraft carrier Langley in January, 1927. Powered by a 525 horsepower water cooled Packard engine and convertable between wheels and floats, these Boeings grossed 3200 pounds and had a high speed of 165-170 mph in landplane configuration.

BOEING FB-6 (USN) In mid-1926 the Navy gave up on the Wright engine installation in the Boeing FB-4 and installed the new Pratt and Whitney R-1340 Wasp radial engine of 400-425 horsepower in the same airframe and redesignated it an FB-6. The success of the combination led to future Navy air cooled engine fighter designs.

BOEING XF2B-1 (MFR.) The next Boeing Navy fighter prototype was the
XF2B-1 shown in company photos of December, 1926 (*top, right*). The XF2B-
1 was a Boeing private venture, Model 89, which was destined to pay off. It
used the Pratt and Whitney Wasp air cooled radial engine and a propeller
spinner along with a split axle landing gear.

BOEING F2B-1 (MFR.) Based on successful testing of the XF2B-1 the Navy
ordered 32 production versions in early 1927 for aircraft carrier use configured as shown in this Boeing photo of September, 1927 (*below*). Changes from
the prototype involved a revised vertical tail and the removal of the large propeller spinner of the prototype. The 2800 pound fighter used the 425 horsepower Wasp engine giving a top speed of 160 mph. Armament included two .30
caliber fixed machine guns and a bomb load capability of 125 pounds. F2B-1
aircraft served on the new carrier Saratoga; it was also the type used by the
Navy's famous flight demonstration team "The Three Seahawks".

BOEING XF3B-1 (MFR.) Another private venture, Model 74, in the Navy fighter field was Boeing's XF3B-1 prototype again using the Wasp engine with F2B-1 wings shown in the company photo of March, 1927. The Navy tested the XF3B-1 both in landplane and in single main float seaplane versions and wanted major changes. The tail stripes shown were unusual on a Navy plane.

BOEING F3B-1 (H.Thorell) With rejection of the XF3B-1 Boeing made major design changes including new wings, completely revised metal tail surfaces, and a revised front fuselage and landing gear. The resulting 156 mph aircraft is shown in the photo with a ring cowl added around the 425 horsepower Pratt and Whitney Wasp engine. After early 1928 flight tests 74 fighters were ordered by the Navy as the F3B-1 with delivery comong later in 1928. These aircraft were soon aboard the new carriers Lexington and Saratoga along with service on the Langley too. They spent about four years as front line fighters and later as utility transports fixed up with cowled engines and wheel pants.

BOEING XF4B-1 (Mfr.) One of two new prototypes provided by Boeing to the Navy for tests is shown at the factory in June of 1928 with a 500 pound bomb under the fuselage and cylinder fairings behind the Wasp engine. Both prototypes were test flown in 1928 as XF4B-1s and were purchased by the Navy after being modified to F4B-1 configuration. These aircraft started a long line of F4B- aircraft, a great favorite of the Navy in the 1930s.

BOEING F4B-1 (MFR.) One of 27 production F4B-1 fighter-bombers in a carrier takeoff shows the pleasing stubby character of the fighter, all of which were delivered in 1929. The Pratt and Whitney Wasp engine was rated at 500 horsepower, span was only 30 feet, normal gross weight 2750 pounds, and maximum speed 176 mph at 6000 feet. Carrier time included service aboard both Lexington and Langley. Some F4B-1s were updated with engine ring cowls and new vertical tails and one was modified into a special F4B-1A VIP transport.

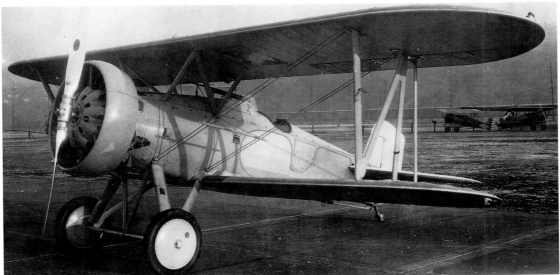

BOEING F4B-2 (MFR.) A 1931 company photo of an F4B-2 on the Boeing hardstand shows it had a ring cowl as standard and a cross-axle landing gear. Another change was incorporation of Frise ailerons. Forty-six planes were procured on a mid-1930 contract with all delivered in the first half of 1931. They were the navalized design opposite to the Army's P-12C fighter. A tail hook and a small tail wheel can barely be seen in the photo.

BOEING F4B-3 (MFR.) The next version of the popular fighter, Boeing Model 235, was the F4B-3. A total of 113 were ordered by the Navy, but only 21 stayed as the -3 version; the remainder being delivered as F4B-4s. The major change was a new all metal fuselage. A rather awkward bomb rack could be carried under the fuselage aft of the landing gear. The photo shows an F4B-3 of the Navy's VF-1 High Hat squadron with a 55 gallon external belly fuel tank attached.

BOEING F4B-4 (MFR.) Doubtless the most famous Navy fighter of the early 1930s, the F4B-4 had a Wasp engine uprated in horsepower and a new vertical tail. A total of 92 -4 versions were delivered mainly in the second half of 1932. Boeing data lists this model as having a span of 30 feet, gross weight of 2900 pounds, top speed of 187 mph, range of 585 miles, and a service ceiling of 27500 feet. This manufacturer's photo shows an F4B-4 ready for the Marine Corps.

BOEING F4B-4 (Author) A restored F4B-4 hangs in the Naval Aviation Hall of the National Air and Space Museum in Washington, DC in early 1981 representing well the biplane fighter era of naval aviation. Boeing built 586 airplanes, including exports, in the F4B-/P-12 series. The last Seattle-built airplane with wooden wings was an F4B-4 delivered on February 28, 1933.

BOEING XF5B-1 (MFR.) An early monoplane Boeing fighter pioneering new all metal semi-monocoque fuselage structure and metal wings was the XF5B-1 or Boeing Model 205, nearly identical to an Army XP-15 prototype. Demonstrated to the Navy in early 1930, the single aircraft was later bought by that service but no production resulted. The Navy was a long way from monoplane production fighters.

BOEING XF5B-1 (MFR.) Another company photo of the XF5B-1 reveals a large number of struts though all strut junctions are carefully faired. Like the F4B-, the XF5B-1 was powered by the 500 horsepower Pratt and Whitney Wasp. A tail hook was fitted for carrier use. Later the propeller spinner was discarded and a ring cowl fitted over the engine.

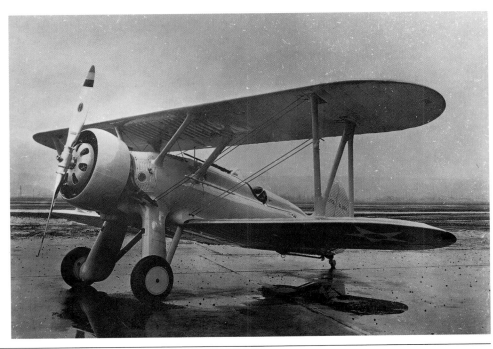

BOEING XF6B-1 (MFR.) This prototype Navy fighter was a Boeing attempt at a biplane successor to its F4B- line. Featuring all metal construction and landing gear strut fairings, the XF6B-1 was ordered in 1931 and delivered to the Navy in the spring of 1933. Powered by a twin row Pratt and Whitney Twin Wasp Junior engine of 625 horsepower, the airplane was tested with both two and three bladed propellers. In 1934 the fighter was redesignated as an XBFB-1 fighter-bomber, but there was no production. Boeing lists wing-span as 28 feet six inches, fighter gross weight at 3700 pounds, and top speed at just 200 mph. The XF6B-1 company photo shown was taken in March of 1933.

BOEING XF7B-1 (MFR.) A September, 1933 company photo shows the original configuration of Boeing's first low wing fighter design for the Navy based on a March, 1933 order for a prototype. After Navy tests in late 1933 there were criticisms including poor pilot vision, high landing speed, and poor maneuverability. Boeing made many design changes including adding flaps, opening the cockpit, and changing the landing gear and engine cowl. The plane was damaged during a high speed dive and a decision was made to abandon the program. Span was 32 feet and gross weight 3500 to 3900 pounds depending on configuration. Speed was about 235 mph using the power of a 550 horsepower Wasp engine.

BOEING XF8B-1 (MFR.) The first Boeing single seater since 1935 was the shipboard fighter-torpedo bomber Model 400, three of which were built as Navy prototype XF8B-1s. First flown in November, 1944, the aircraft was meant for attacking Japan from the larger carriers. Powered by a Pratt and Whitney 3000 horsepower Wasp Major engine driving a 13.5 foot diameter Aeroproducts dual rotation propeller, the 54 foot span 20500 pound gross weight XF8B-1 had a high speed of 432 mph at 27000 feet and could carry two 1600 pound bombs internally. Armament consisted of six wing guns. With war's end and the jets coming the XF8B-1 program was shortly dropped.

CURTISS TS-1 (MFR.) Shown in twin float seaplane form in a Curtiss photo, the TS-1 was a Naval Aircraft Factory single seat fighter design powered with a 200 horsepower Lawrence J-1 air cooled radial engine, the fore-runner of a Navy trend. Thirty-four were ordered from Curtiss in late 1922 and four more were built at the NAF. Wheels and floats were interchangeable. The TS-1s had a high speed of about 124 mph, landplane or seaplane. They served through the 1920s including as seaplanes aboard battleships. Some TS-1s were revised with other engine installations.

NAVAL AIRCRAFT FACTORY TR-3 (USN) A racing version of the TS-1 with twin floats and re-engined with a Wright E-2 Hisso water cooled model of 180 horsepower, this TR-3 was entered in the 1922 Curtiss Marine Trophy Race but did not finish. The seaplane was later reworked as a very streamlined TR-3A racer.

CURTISS F4C-1 (MFR.) September, 1924 company photos (*above*, *left*) of the F4C-1 show a design based on the TS-1 with the major changes of an all metal aluminum tube fuselage and a metal wing structure along with a lower wing attached directly to the fuselage and a Wright J-3 engine of 200 horsepower. Although there was no production the Navy learned there were considerable advantages to metal construction. The F4C-1 weighed about 200 pounds less than the TS-1 for one thing.

CURTISS F6C-1 (MFR.) The Navy version of the Army P-1 Hawk, the F6C-1 (*at right, right center*) differed only in Navy equipment items. Nine were ordered and first flight took place in August, 1925. The manufacturer's photo shows an F6C-1 tested as a twin float seaplane the next month. Four of the planes were revised for carrier operation and labeled F6C-2s. Powered with Curtiss D-12 engines of 435 horsepower, the 2800 pound gross weight fighter had a high speed of slightly over 160 mph.

CURTISS F6C-3 (M.Sheppard) The next Curtiss Navy Hawk fighter was the F6C-3 (*below*), 35 of which were split between Navy and Marine Corps in 1927. These aircraft were also convertable to twin floats and some were so used for a period. The -3 aircraft were a little slower and slightly heavier than the -1s and were generally similar to the Army P-1A. The photo shows a Marine F6C-3 at an early airshow.

CURTISS F6C-4 (USN) The Navy was converting to air cooled radial engines and like the Boeing fighters the Hawk was revised in September, 1926 to use a Pratt and Whitney Wasp of 410 horsepower using the first F6C-1 for a prototype. Thirty-one F6C-4s were delivered in early 1927 and the Marines got a few.

CURTISS F6C-4 (Wright Aero.) One of the F6C-4 aircraft, number A-7403, was used as a powerplant test bed for various installations. The photo shows this airplane with its radial engine in a quite modern looking cowling along with a forward fuselage revised to conform to cowl lines. Service aircraft did not have this modification.

CURTISS F6C-5 (USN) A Navy photo shows the XF6C-5 version of the Hawk, a single aircraft which was the first F6C-1 initially revised to an F6C-4 prototype and later modified to use a Pratt and Whitney R-1690 Hornet engine installation of 525 horsepower as the XF6C-5, the fore-runner of many other Navy Hornet installations. Gross weight was about 3000 pounds and high speed almost 160 mph.

CURTISS XF6C-6 (MFR.) A streamlined high wing monoplane racer converted from an F6C-3 fighter (*above, right*), and sometimes called the Page Racer, was designated an XF6C-6 but was not a fighter at all. Groomed for entry into the 1930 National Air Races, the racer was powered by a version of the 600 horsepower Curtiss V-1570 Conqueror engine. The racer crashed at the NAR in Chicago where it was believed that Marine Capt. Page passed out from inhaling engine exhaust fumes.

CURTISS XF6C-7 (USN) The F6C-4 number A-7403 was again used for an engine test bed, this time under the XF6C-7 designation. In the shops of the Naval Aircraft Factory in Philadelphia the airplane was used for experimental installation of a Ranger V-770 engine of 450 horsepower, Navy interest no doubt provoked by the air cooled feature of the engine. No production resulted from the testing however.

CURTISS XF7C-1 (MFR.) The XF7C-1 Seahawk was a new carrier fighter design by Curtiss first flown in February, 1927. Like previous designs it could be operated as a seaplane, in this case with a single main and two tip floats. Powered with a Pratt and Whitney R-1340 Wasp engine of 450 horsepower the prototype was distinguished by a large propeller spinner up front.

CURTISS F7C-1 (MFR.) A November, 1928 photo (*right*) of a "production" F7C-1 Seahawk at the Curtiss plant shows the upper wing outer panel sweepback and an external belly fuel tank. A total of 17 Seahawks were produced in late 1928. The large spinner was omitted and the landing gear revised. The F7C-1 was used by the Marines and did not fulfil its intended purpose as a carrier or battleship fighter.

CURTISS F7C-1 (MFR.) The F7C-1 Seahawk fighter (*below*) was used for certain experiments such as a twin tandem propeller installation (not shown) and, as shown in the photo, exploration of the effects of wing leading edge automatic slats. When they popped out the wing lift curve extended such that the airplane could operate at a much higher angle of attack.

CURTISS F8C-1 (USN) After considerable confusion over original June and November, 1927 Navy-Curtiss contracts, the impetus of a Nicaragua conflict and US Marine Corps operations in China resulted in revised 1928 contracts providing delivery to the Marines of six two seat F8C-1s like Army O-1B Falcons and 21 similar F8C-3s like Army A-3 attack planes except for Navy use of the Wasp radial engine. The F8C-1 later became an OC-1 observation and the F8C-3 an OC-2. The Navy photo shows a 430 horsepower Wasp powered F8C-1/OC-1 in March of 1928. These aircraft were really observation types. Gross weight was about two tons and high speed about 135 mph. The Marines used Falcons in the Nicaraguan fighting.

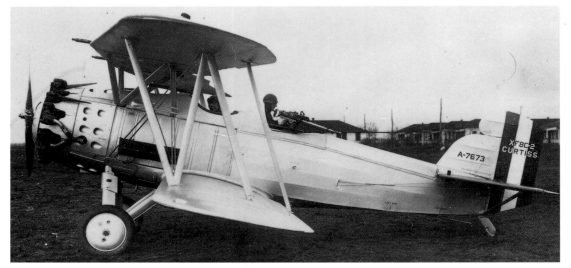

CURTISS XF8C-2 (MFR.) The XF8C-2 shown in this November, 1928 photo was classified as a two seat fighter and dive bomber, and was a much different design than the F8C-1 and F8C-3, though powered with the same Wasp engine of 450 horsepower. Ordered in March of 1928 the first aircraft crashed and a replacement was ordered.

CURTISS XF8C-2 (MFR.) An April, 1929 company shot of the second XF8C-2 shows a new long chord engine cowling around the Wasp engine presaging designs of the future. Armament was two .30 caliber fixed guns in the upper wing and two flexible guns in the rear cockpit. Testing of the two XF8C-2 prototypes led to a production order for a similar F8C-4 Helldiver. High speed of the 3400 pound aircraft was 145 mph.

CURTISS XF8C-4 (MFR.) The single prototype of the production Helldiver, as the new two seat carrier fighter was called, is shown in a run-up shot at Curtiss. Shown uncowled in this photo, the 450 horsepower Pratt and Whitney Wasp engine sometimes had a narrow chord ring cowl around it. The single XF8C-4 flew in the spring of 1930 and successful tests resulted in orders for production F8C-4s.

CURTISS F8C-4 (Thorell) A photo of 1930 shows a production F8C-4 Helldiver, one of 28 ordered the year before. The two seat fighter soon joined the fleet. Slower and less maneuverable than the current single seaters, these airplanes were soon redesignated as O2C-1 observation planes and went back ashore. The 450 horsepower Wasp powered the Helldiver to a top speed of about 140 mph. Wing span was 32 feet and gross weight was approximately two tons.

CURTISS F8C-5 (USN) Another version of the Curtiss two seat Helldiver was for the Marines and was minus carrier gear. Though given a fighter designation the more than 60 new aircraft were soon changed to O2C-1 observation Helldivers at Marine land bases in 1931. The Navy photo of an F8C-5 with its two fixed upper wing guns and two flexible guns aft was taken in September, 1930.

CURTISS F8C-7 (MFR.) One of four Helldivers equipped with 575 horsepower Wright Cyclone engines, the F8C-7 VIP transport was really not a fighter. The company photo taken in December, 1930 shows a Helldiver with wheel pants (including the tail wheel!), strut and wire "cuff" fairings, engine ring cowl, and enclosed cockpits. Curtiss called it a "Cyclone Command" aircraft.

CURTISS F8C-7 (MFR.) Another view of the special F8C-7 Cyclone Command aircraft in December, 1930 shows the special attention paid to streamlining. The aircraft had no armament; upper wing gun ports are taped over. High speed was about 180 mph low down and weight was a couple of hundred pounds greater than a standard Helldiver even without armament.

CURTISS XF8C-7 (MFR.) Another of the four Cyclone powered Curtiss Helldivers, the XF8C-7 started as a company private venture as indicated by the civil registration on the vertical tail. Also photographed in December, 1930 (four days after the F8C-7 above), the XF8C-7 looked a bit more warlike with armament installed along with ring engine cowl, wheel pants and cockpit enclosure. It was later purchased by the Navy and labeled, along with the other Cyclone Helldivers, as an O2C-2.

CURTISS XF9C-1 (MFR.) Ordered as a small fighter with a 25 foot six inch span based on a Navy design, and competitive with the Fokker XFA-1 prototype fighter, the Curtiss XF9C-1 flew first in February, 1931 as a carrier type as shown in the photo. Powered with an R-975 Wright Whirlwind engine of 420 horsepower, the little Sparrowhawk weighted only 2500 pounds gross and had a top speed of about 175 mph. Later the XF9C-1 pioneered as an airship hook-on fighter tested ahead of the F9C-2 aircraft.

CURTISS F9C-2 (MFR.) Based on success of the XF9C-1 six F9C-2 Sparrowhawk fighters (sometimes called the Akron fighter) were ordered for airship-based duty on Akron and Macon with a hook device designed to catch on a lowered airship trapeze as shown in the photo. The first F9C-2 flew in the spring of 1932. Note the airplane still has a carrier tail hook, and F9C-2s sometimes did operate off a carrier.

CURTISS F9C-2 (C.Mandrake) Another view of the F9C-2 airship fighter, still with a carrier tail hook. Blast tubes for two .30 caliber guns just nosed between engine cylinders and got inside the engine ring cowl. Note the panted streamlined landing gear, the fact that the upper wing connected directly to the fuselage, and the aircraft upper hook support struts. The -2 was a little heavier than the -1, but performance was similar.

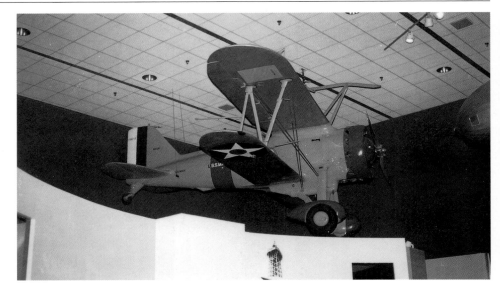

CURTISS F9C-2 (Author) A photo of an F9C-2 as it hangs in the National Air and Space Museum in Washington, DC. (with airship hook but no carrier hook.) The F9C-2 performed operational search missions from the airships, sometimes without their landing gear, but when the Akron and Macon went down in 1933 and 1935 most of the Sparrowhawks were also lost.

CURTISS XF9C-2 (MFR.) A late surviving Sparrowhawk which started as a Curtiss demonstrator plane is shown in a company photo of July, 18, 1932. This was the aircraft Curtiss exhibited at aeronautical shows. Used as a test plane, in this case for an early Curtiss two bladed controllable pitch Electric propeller, the XF9C-2 also had a much revised wire braced single leg landing gear. Note the civil registration on the tail.

CURTISS XF11C-1 (MFR.) There was no XF10C-1 from Curtiss though one of the Cyclone powered two seat Helldivers (No.8847) was considered for such a designation at one time it never got on an airplane tail. The XF11C-1, shown in a Curtiss photo of September, 1933 was a new Navy Hawk type single seat fighter-bomber with a new Wright R-1510 two row engine of 600 horsepower, a three blade propeller, and large panted wheels. High speed was just over 200 mph. Later called a BFC-1, this aircraft never reached production and was used for powerplant tests by the NASA.

CURTISS XF11C-1 and F11C-2 (MFR.) A company flight photograph of November, 1932 shows the twin row R-1510 engined XF11C-1 in the foreground and an F11C-2 in the background using a single row nine cylinder Wright Cyclone. The Cyclone powered Hawk got the nod for a production contract with the single XF11C-1 used only for tests.

CURTISS XF11C-2 (MFR.) Built as a Curtiss company demonstrator, the XF11C-2 differed from the -1 in engine type, being powered by a 700 horsepower Cyclone, longer landing gear, and smaller low pressure tires, along with a forward located tail wheel as can be seen in the company photo of March, 1932. The belly external fuel tank was flattened like those of export Hawk II models, of which there were many. Tested by the Navy in May, 1932 after being purchased in April, the XF11C-2 led directly to a Goshawk production contract.

CURTISS F11C-2 (USN) A fine USN photo shows an F11C-2 in production configuration. The tail wheel has been moved aft as on the XF11C-1. 28 F11C-2 Goshawks were ordered by the Navy, one of which was held back as an XF11C-3. The F11C-2 could carry a 500 pound bomb under the fuselage and smaller bombs under the wings; two cowl machine guns were mounted. The planes went to carrier Saratoga in early 1933. In 1934 these planes were slightly modified and redesignated as BFC-2 fighter bombers.

CURTISS XF11C-3 (MFR.) One of the production F11C-2 aircraft was revised at the factory to employ retractable landing gear by modifying forward fuselage lines to enclose the mechanism and retracted wheels. The company photo shows a wheels-retracted view of the XF11C-3. A 700 horsepower Wright Cyclone powered the 4500 pound fighter to a high speed of about 230 mph, an approximately 25 mph increase over a fixed gear F11C-2. The production F11C-3s quickly became BF2C-1 bomber-fighters in 1934.

CURTISS XF12C-1 (MFR.) A July, 1933 company photo shows the two seat high wing strut braced all metal XF12C-1 fighter, certainly a novelty! The wing spanned 41 feet six inches and was manually foldable aft. A tail hook was fitted to this carrier fighter candidate. The Navy was still ambivalent about two seat fighters; this one had a 624 horsepower Wright R-1510 engine and a top speed of about 215 mph. The Navy later redesignated the plane as a scout and then a scout-bomber, but there was no production.

CURTISS XF12C-1 (MFR.) An unusual flight photo of the XF12C-1 after being re-engined with a Wright R-1820 single row nine cylinder Cyclone shows the many struts required to support the big parasol wing. Blast tubes for cowl guns are above and behind the engine cowl. Slats and flaps adorned the high wing. For awhile the plane became an XS4C-1 scout and finally in 1934 was redesignated an XSBC-1 scout bomber.

CURTISS XF13C-1 (MFR.) A December, 1933 photo of the Curtiss XF13C-1 single seat monoplane disclosed a clean configuration with the wing cabane struts integrated into the pilot's cabin. The plane was tested by the Navy in early 1934. Powered by a Wright XR-1510 two row engine of 700 horsepower, this 2.2 ton gross weight twin gun all metal monoplane had a high speed of about 235 mph. The wing included slats and flaps.

CURTISS XF13C-1 (NACA) With the number of wing support struts minimized compared to the XF12C-1 because of the pilot's cabin arrangement, the XF13C-1 looked relatively clean in this flight view. Interestingly though reports have the plane tested aboard a carrier the photos show no tail hook installed.

CURTISS XF13C-2 (MFR.) The biplane version of the XF13C-1 is pictured in a December, 1933 photo with new wings installed on the same airplane. The designation on the tail was not even changed though the plane is now an XF13C-2. There was only one airplane; there was a bad depression in the country. The biplane was about 15 mph slower and was slightly lighter. The engine was the same. At this time the Navy was loath to give up the biplane arrangement; it took a few more years.

CURTISS XF13C-3 (C.Mandrake) The Navy kept testing the XF13C- airplane with changes to the tail now made making the same aircraft an XF13C-3 monoplane. It was turned over to the NACA as a test aircraft and later, in 1937, went to the Marines as a utility plane. The new tail now showed the Xf13C-3 designation. In any case the Navy was not interested in the aircraft as a production fighter.

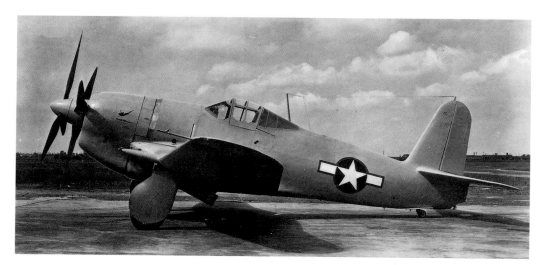

CURTISS XF14C-2 (MFR.) Curtiss garnered another Navy contract in 1941 for an XF14C-1 high altitude fighter using a new liquid cooled Lycoming engine as the XF14C-1. It became clear by 1943 the engine program held no promise, so Curtiss designed an XF14C-2 (*left, below*) using a turbosupercharged Wright R-3350 engine of 2200 horsepower at high altitude driving a dual rotation propeller as shown in the company photo of September, 1943. Tests began shortly afterwards and the Navy took delivery in July, 1944. Performance of the seven ton fighter was disappointing, and the Navy at that time had no need for such a fighter.

CURTISS XF15C-1 (MFR.) A July, 1945 company photo shows the initial configuration of the XF15C-1, a composite-powered fighter with a Pratt and Whitney Double Wasp of 2100 horsepower up front and a 2700 pound thrust Allis Chalmers J36 jet engine aft. The Navy provided a contract for three aircraft in April, 1944. Initial flight (without the jet engine) took place in February of 1945 and the jet was installed later in May just before the first aircraft crashed. The initial version had a low horizontal tail.

CURTISS XF15C-1 (MFR.) The second and third prototypes resumed tests with a new T-tail in 1946 as shown in the photo illustrating the XF15C-1 cruising on the jet engine only with the propeller feathered, a rather impractical cruise method. Wing span was 48 feet, normal gross weight 16600 pounds, and the dash speed attained using both engines was about 460 mph. After the Navy decision to equip with all jet fighters the composite power idea died and no Curtiss production resulted. The XF15C-1 was the last Curtiss Navy fighter.

DOUGLAS XFD-1 (Thorell) The Navy could not seem to get away from a two seat fighter concept, and in June of 1932 ordered an XFD-1 from Douglas along with a Vought XF3U-1 and a Curtiss XF12C-1 at the same time. The XFD-1 was a clean cut 4750 pound gross weight biplane powered by a 700 horsepower Pratt and Whitney R-1535 engine providing a high speed of just over 200 mph. Two fixed forward firing .30 caliber guns in the fuselage were supplemented by a rear flexible gun. There was no production.

DOUGLAS XF3D-1 (USN) Ordered by the Navy in April, 1946 the three two place twin jet engine Douglas XF3D-1s were night fighters with nose radar and four fuselage mounted 20mm cannon. Certainly the most sophisticated night fighter of its time, the Skyknight was powered by Westinghouse J34 jet engines each of 3000 pounds jet thrust. A mid-wing monoplane with side-by-side seating for two, the aircraft could carry bombs or drop tanks underwing. The photo of a prototype XF3D-1 was taken in March of 1950.

DOUGLAS F3D-1 (P.Bowers) The photo shows one of 28 first production F3D-1 Skyknights under test at NATC Patuxent River, Md. in 1950; the F3D-1 first flew in February of that year. Underpowered with two 3250 pound thrust Westinghouse jets, the 24500 pound Skyknight had a high speed of about 475 mph at altitude. The F3D-1s were used as development aircraft for later models. Most were used as USMC trainers.

DOUGLAS F3D-2 (MFR.) The major production version of the Skyknight night fighter was the F3D-2 with uprated 3400 pound thrust J34 jets since the 4600 pound thrust J46 engines were not ready. A total of 191 F3D-2s were built in the early 1950s after a February, 1951 first flight. The aircraft were land-based and used by the Marines. A small quantity were used in Korea and achieved a few victories.

DOUGLAS EF-10B (R.Besecker)
The Skyknight had a long life as evidenced by this May, 1969 photo of an EF-10B, the redesignation in 1962 of 30 F3D-2Q Marine radar countermeasures versions. About a year later the USMC finally got the aircraft replaced. Three hundred gallon drop tanks are on the stores pylons. Wing fold was simple; though carrier-hooked, these aircraft normally stayed ashore.

DOUGLAS F3D-2 (AAHS) Along with 16 Skyknights delivered as F3D-2Q missile firing aircraft, some F3D-2s were modified as F3D-2T trainers like the one shown in the photo belonging to Navy all weather fighter squadron VFAW-3. With side-by-side seating the Skyknight was very useful as a trainer. Note the tail hook and tail bumper, the latter on all Skyknight aircraft.

DOUGLAS XF4D-1 (USN) The result of a 1947 Navy request to Douglas for design of a short range fast-climbing interceptor, the Skyray first flew in January, 1951 with a substitute Allison J33 engine of 5000 pounds thrust; later a 7000 pound thrust Westinghouse J40 with an afterburner was installed. Using the latter engine the Skyray set a world speed record of 753 mph in October, 1953. The photo shows an XF4D-1 that same month coming in to land on CVA-43, the USS Coral Sea.

DOUGLAS F4D-1 (USN) A major redesign of the Skyray resulted in the production F4D-1 with a Pratt and Whitney afterburning J57 engine of 10000 pounds thrust (16000 with afterburner) and first flight was in June of 1954. High speed was well over 700 mph low down. Climb of the "FORD" was spectacular and many records were set; one was almost to 40000 feet in two minutes! Standard armament was four 20mm cannon; unguided rockets could also be fitted. The photo shows a Patuxent River test F4D-1 on a CVA-59 elevator in April of 1956 and illustrates the planform and method of folding.

DOUGLAS F4D-1 (H.Andrews) An F4D-1 Skyray of Navy Squadron VF-101, the Grim Reapers, on board CVA-62, the USS Independence. One of many squadrons with the Skyray, the Reapers used the fighter from 1956 to 1962, then transitioning to the F-4 Phantom. A total of 419 Skyrays were produced, the last aircraft coming off the line near the end of 1958. In 1962 the Douglas was redesignated to an F-6A.

DOUGLAS F4D-1/F-6A (USN) A Skyray of Navy Squadron VU-3 (UTRON-3) which controlled drone aircraft for fleet target practice. The F-6As were revised to provide airborne drone control capability and were actually designated DF-6As. The squadron started operating Skyrays from early 1962 and was the last Navy operational squadron to use them. All were retired by October, 1964. There was a great deal of criticism about Skyray flying qualities over its lifetime.

DOUGLAS XF5D-1 (MFR.) Originally an F4D-2, the XF5D-1 Skylancer was a higher speed development of the F4D-1. First flown in April of 1956 with a Pratt and Whitney J57 engine providing a high speed of almost 800 mph at altitude, and designed for missile armament, the F5D-1 was supposed to have a more powerful General Electric J79 engine which was never installed. Possibly based on the mixed review reputation of the F4D-1 the Navy decided to cancel the Skylancer effort. Four aircraft were completed out of a planned eleven in a pre-production lot.

GRUMMAN XFF-1 (P.Bowers) Conceived in 1930, the XFF-1 was the first total airplane product of Grumman. A two seat biplane fighter with enclosed cockpits, the big innovation was a retractable landing gear. First flight of this all metal framed fighter took place in December of 1931. Flight testing was successful and the Wright Cyclone model engine finally installed got high speed just over 200 mph, making it the fastest Navy fighter of that time.

GRUMMAN XGG-1 (P.Bowers) Though not a Navy aircraft, the XGG-1 was so representative of the early FF-1/SF-1 aircraft it is included. Built as a company demonstrator and first flown in late September, 1934, the plane was originally powered with a Pratt and Whitney R-1340 Wasp of 450 horsepower. It had a gross weight of 3640 pounds and a maximum speed of 206 mph. It was sold to Canadian Car and Foundry Ltd. in 1936 and was a prototype for the Canadian Goblin.

GRUMMAN FF-1 (USN) A production order for 27 FF-1 two seat fighters was given to Grumman and the first flight occurred, along with the first delivery, on April 24, 1933 with the final FF-1 turned over in November of that year. With a 750 horsepower Wright Cyclone F, three .30 caliber machine guns, and capacity for two 100 pound bombs, the 4650 pound FF-1s had a top speed of about 210 mph. They were first assigned to Navy squadron VF-5S. One of that squadron's planes is shown in the photo.

GRUMMAN FF-2 (H.Thorell) The FF-2 was a fighter-trainer version of the FF-1, in fact the Naval Aircraft Factory converted 22 of the FF-1s to dual controls. Apparently three of the aircraft had crashed because only the 22 are shown as FF-2 versions. One of these is shown running up its big Wright Cyclone radial and two blade 9 foot six inch diameter propeller.

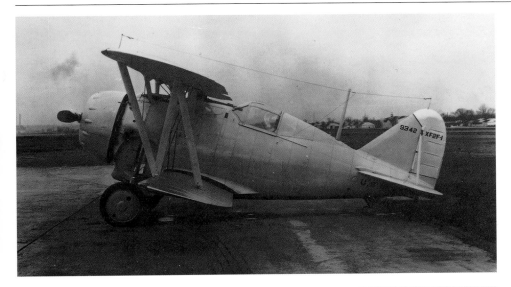

Grumman XF2F-1 (USN) The first single seat Grumman fighter prototype is shown in a Navy photo of January, 1934, almost exactly three months after Grumman test pilot Hovgard took it up for a first flight. Ordered in November, 1932, it was delivered to the Navy on October 18, 1933. Powered by a Pratt and Whitney R-1535 Twin Wasp Junior of 625 horsepower this stubby little biplane grossed at 3500 pounds and attained a 240 mph top speed. Wingspan was only 28 feet and length 21 feet three inches. Testing was sufficiently successful to warrant a production order.

GRUMMAN F2F-1 (USN) Grumman hit it big with a production order for 55 F2F-1 fighters; that was pretty large for those days. Specifications were similar to the prototype. First flight took place January 9, 1935 and deliveries went on between mid-January to early August of that year. Armament was two .30 caliber fuselage guns and two 100 pound bombs. Two of the planes went to the Marine Corps; the remainder went to Navy Squadrons VF-2 and VF-3 initially. The July 1938 flight photo of the stubby craft shows it belongs to VF-5. The F2F-1 was still being used in 1940.

GRUMMAN XF3F-1 (AAHS) A larger, heavier, and somewhat more powerful version of the F2F-1 was Grumman's XF3F-1 ordered in late 1934. The XF3F-1 ran into trouble during dive testing; three prototypes were built. The first crashed March 22, 1935; the second crashed that May, but the third was revised in three weeks and passed all tests. It was delivered in mid-June of 1935. With a 700 horsepower R-1535 engine the plane had a high speed of about 225 mph.

GRUMMAN F3F-1 (USN) Another large (for the time) production order for the F3F-1 fighter was given Grumman and a first flight piloted by Lee Gelbach took place in late January, 1936. All 54 aircraft were delivered by mid-September of that year. Production aircraft went to Navy Squadrons VF-4 and VF-6. The Twin Wasp Junior engine powered the 4100 pound plane to a maximum speed of 230 mph. The Navy photo of April, 1940 (*above*) shows an F3F-1 of VF-7 with the star insignia representing the Neutrality Patrol.

GRUMMAN F3F-1 (USN) A flight view of an F3F-1 of Squadron VF-4 taken on January 17, 1939 shows the configuration of this compact pre-war biplane splendidly. The cowl was wrapped tightly around the R-1535 engine. The propeller was a Hamilton Standard eight foot six inch hydraulic controllable pitch model. The F3F-1 was Grumman Company design Model 11.

GRUMMAN XF3F-2 (USN) A revised Grumman fighter, heavier and more powerful, was flown first on July 21, 1936 and delivered to the USN as the XF3F-2 six days later. Major changes were use of an 850 horsepower Wright XR-1820 Cyclone nine cylinder single row engine in place of the two row R-1535, a new three blade Hamilton Standard nine foot diameter constant speed propeller, a revised rudder with more area, and one .50 caliber and one .30 caliber guns along with the two hundred pound bombs inderwing. The photo shows the aircraft in carrier landing configuration with hook down.

GRUMMAN F3F-2 (Thorell) The largest Navy fighter contract to date was gained by Grumman with Navy purchase of 81 production F3F-2 aircraft the first of which was delivered on the same day of its first flight, July 27, 1937. The last was delivered in May of 1938. The 850 horsepower Cyclone produced a maximum speed of 260 mph from the 4500 pound fighter. The F3F-2s were delivered to Navy Squadron VF-5 and Marine VMF-1 and VMF-2. These planes were used as trainers in early World War II.

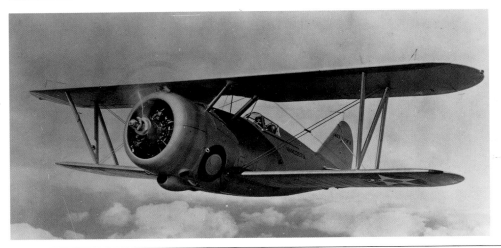

GRUMMAN F3F-3 (USN) The final version of the Grumman biplane Navy fighter after test of an F3F-2 Number 1301 to F3F-3 configuration with a curved windshield and revised cowling along with minor landing gear changes. First flown on December 15, 1938 and delivered the next day, all 27 were put in Navy hands by May, 1939. The fastest biplane at about 275 mph, it went to Navy VF-5 and Marine Corps VMF-1. Some of these planes went aboard carrier Enterprise after Pearl Harbor! The photo shows an F3F-3 during tests at Anacostia in April of 1939.

GRUMMAN XF4F-2 (MFR.) The initial monoplane Grumman Navy fighter was a replacement for an intended biplane XF4F-1 in 1936. First flying in early September, 1937 and delivered to the Navy in December, the XF4F-2 was powered by a Pratt and Whitney R-1830 Twin Wasp of 900 horsepower. With a span of 34 feet and gross weight of 5330 pounds this prototype had a maximum speed of 290 mph. The XF4F-2 had a crackup in April, 1938 and was rebuilt with many changes as the XF4F-3. Grumman lost the production contract to Brewster and its XF2A-1.

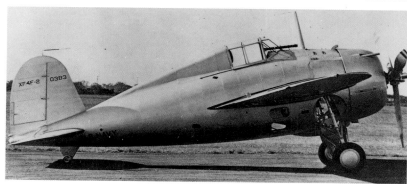

GRUMMAN XF4F-3 (USN) A Navy photo shows the rebuilt airplane as the XF4F-3 in August of 1939. Two major changes were a new larger area wing spanning 38 feet and a new version of the Twin Wasp with a two stage mechanical supercharger. Flown first in February, 1939, the plane was delivered March 7, 1939. Another change was a four gun armament with two .30 caliber guns in the cowl and two .50 caliber guns in the wings. Many tests were made with propeller spinners, varying horizontal tail locations, and vertical tail changes. The plane crashed at NAS Anacostia. Loaded weight came to 6300 pounds and maximum speed was recorded at 333 mph.

GRUMMAN F4F-3 (USN) The first Wildcat model in World War II action, the F4F-3 was flown initially by Grumman test pilot Selden Converse on February 24, 1940 with first delivery to the Navy in late November of that year. A total of 285 F4F-3 models were delivered, the last 100 being extra planes for training. The initial 36 airplanes were delivered in the old-style bright Navy colors; the rest were in gray. The Wake Island Wildcats were -3 models.

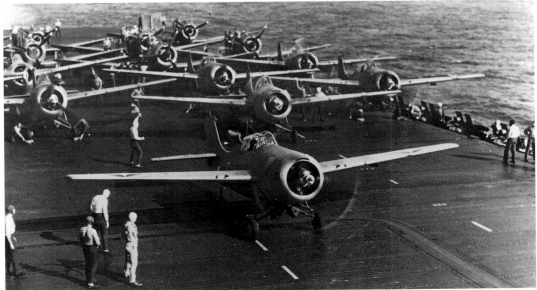

GRUMMAN F4F-3 (MFR.) A flight of early F4F-3 Wildcats painted up for pre-war maneuvers is shown. Powered by a 1200 horsepower two stage supercharged Pratt and Whitney R-1830 Twin Wasp engine the 7060 pound normal gross weight -3 Wildcat had a high speed of about 330 mph at medium altitude. Four .50 caliber machine guns were mounted in the fixed wings and two 100 pound bombs could be carried.

GRUMMAN F4F-3 (USN) A duo of very early F4F-3 fighters are shown in a Navy photo. The Navy had to scramble to get protected fuel tanks into the fighters. F4F-3 aircraft were supplied to Navy Squadrons VF-4, VF-42, VF-72, and Marine Squadrons VMF-111 and VMF-211. One F4F-3 was tested on twin floats as the F4F-3S "Wildcatfish" but no more seaplane versions were produced.

GRUMMAN F4F-3A (USN) The F4F-3A Wildcat used a Twin Wasp engine with only a single stage supercharger. A total of 95 were built, the first 30 exported for Greece but taken over by the British as the Martlet III. All were delivered from Grumman in the period March through May, 1941. The prototype for the F4F-3A was the single XF4F-6 which crashed in May, 1942. The first F4F-3A delivery to a US unit was for Marine Squadron VMF-1 in April, 1941. High speed was 312 mph at a medium altitude.

GRUMMAN F4F-4 (MFR.) The F4F-4 Wildcat was the major production model. Important changes were provision for manual wing folding and a six gun armament (with less ammunition per gun) in place of four. First flown in November, 1941, a total of 1169 F4F-4 Wildcats were delivered to the Navy from November 25, 1941 until the end of 1942. The -4 ended up 1100 pounds heavier than the -3A and was considered sluggish by its pilots.

GRUMMAN F4F-4 (USN) A new F4F-4 Wildcat of Navy Squadron VF-41 poses for the photographer with the early striped rudder and the red center circle still on the fuselage side national insignia. The engine was a Twin Wasp; the propeller was a three bladed Curtiss Electric model of nine feet nine inches diameter. The Wildcat had to hold off the Japanese until newer Navy fighters got into the war. High speed was just under 320 mph.

GRUMMAN F4F-4 (USN) An F4F-4 Wildcat with tail hook out catching a wire lands on a carrier. The aircraft had good carrier landing characteristics, but the hand-cranked retractable landing gear gave some pilots trouble. When Grumman stopped Wildcat production Eastern Aircraft Division of General Motors turned out 1150 more FM-1 near-duplicates with only four guns.

GRUMMAN XF5F-1 (MFR.) Shown in a Grumman photo, the single XF5F-1 twin engined Skyrocket fighter was first flown April 1, 1940, but was not delivered to the Navy until February of the next year. Planned nose armament was four 23mm cannon, but these were unavailable and machine guns were substituted. Compartments for ten anti-aircraft "bomblets" were provided in the foldable outer wing panels. Span was 42 feet, and two 1200 horsepower Wright Cyclones powered the Skyrocket to a maximum speed of about 380 mph. Gross weight was 10900 pounds.

GRUMMAN XF5F-1 (MFR.) The XF5F-1 Sky-rocket in final configuration during a test flight. Here counter-rotating engines and propellers were installed. Following initial tests the fighter was returned to Grumman in April, 1941 for major modifications. It was damaged in 1942 and refurbished and later damaged again. It was off the Navy inventory in December, 1944. No production was gained though the XF5F-1 was a design aid in later F7F-1 development.

GRUMMAN XF6F-1 (MFR.) Two prototypes of a new XF6F-1 Navy fighter to be developed as a backup for the Corsair were ordered from Grumman in mid-1941. First flight was in late June, 1942 powered by a Wright R-2600 engine with a 13 foot diameter Curtiss propeller. Gross weight was 11630 pounds and high speed 375 mph. This plane was later modified with a Pratt and Whitney R-2800 Double Wasp engine in August, 1942 and then shortly turned into the second XF6F-3.

GRUMMAN XF6F-3 (MFR.) Prototype for the F6F-3 Hellcat shows in a company photograph using a Pratt and Whitney R-2800 Double Wasp with a two stage two speed supercharger. First flight was July 30, 1942. It was eventually given to the Navy in late 1943, but earlier in mid-August it was badly damaged in a forced landing after engine failure. This prototype was repaired and later became the XF6F-4 used for armament testing.

GRUMMAN F6F-3 (MFR.) A Grumman photo shows two new F6F-3 Hellcat fighters in flight. This version was the first production model; the propeller spinner was eliminated and the landing gear fairings simplified. A large fighter with a 42 foot ten inch wingspan and a gross weight of about 12450 pounds, the Hellcat development was on a fast track and it first got into action the end of August, 1943. Armament consisted of six .50 caliber wing guns.

GRUMMAN F6F-3 (R.Besecker) A post-war civilian F6F-3 Wildcat taxis out for a flight at a July, 1961 airshow. Though at about 375 mph high speed it was substantially slower than the Corsair, the Hellcat was a fine performer against Japanese aircraft and adapted very well to carrier decks. A total of 4402 F6F-3 Hellcats were produced with many going to the British under Lend-Lease. Two 1000 pound bombs could be carried or one torpedo.

GRUMMAN F6F-3N (MFR.) The Hellcat was a favorite night-fighter. The photo shows an F6F-3N night fighter pictured in August, 1945. A total of 205 aircraft were converted from -3s with APS-6 radar with a radome on the right wing. Also shown on this aircraft is a "mixed battery" of armament (four .50 caliber machine guns and two 20mm cannon). Cockpit changes were also made for night fighting.

GRUMMAN F6F-5 (R.Besecker) The most-produced Hellcat with 7870 aircraft built was the F6F-5 (*above, right*). First flight was on April 4, 1944 with a first delivery 25 days later and these continued until November 16. 1945. Changes from the -3 model included an improved windshield and canopy, strengthened tail, spring tabs on the ailerons, a new pinched-in cowl, and a new extra-smooth paint finish. Most aircraft had six .50 caliber machine guns; some had a "mixed battery" of four .50 calibers and two 20mm cannon. All aircraft had water injection systems on their 2000 horsepower Pratt and Whitney Double Wasp engines. High speed was about 380 mph; gross weight about 12600 pounds. One photo shows a post-war F6F-5 from a reserve unit.

GRUMMAN F6F-5K (P.Bowers) The F6F-5K was a converted F6F-5 equipped as a drone aircraft stripped of armament and having special electronics installed to receive signals from a controller aircraft. A few were used in Korea against tough targets starting in August of 1952. Launched from aircraft carriers and filled with explosives, they were guided by a mother aircraft such as a Skyraider or an F6F-5D controller plane.

GRUMMAN F6F-5N (R. Besecker) Conversions of -5 airplanes, the F6F-5N models were the most prolific night fighters with 1434 so modified, including 80 for the British. Equipped with an APS-6 radar pod on the right wing, these aircraft replaced Army P-61 night fighters in the Philippines fighting for a time. High speed was about 370 mph and gross weight was approximately 13000 pounds.

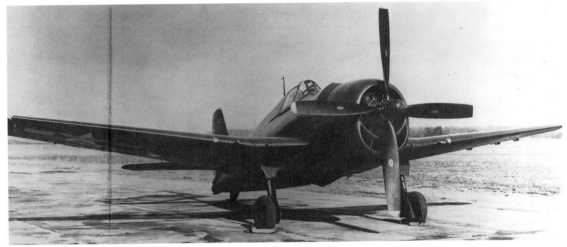

GRUMMAN XF6F-6 (USN) A Navy photo dated December 20, 1944 shows the last Hellcat type, an XF6F-6, equipped with an uprated 2100 horsepower water-injected Double Wasp C engine which could put out 2540 horsepower using water injection and driving a four blade 13 feet two inch diameter Hamilton Standard propeller. Only two prototypes were made, revised from -5 airframes. First flight took place on July 6, 1944 with delivery the same day. No production resulted. The XF6F-6 was the fastest Hellcat at about 415 mph; gross weight was 12800 pounds.

GRUMMAN XF7F-1 (USN) The first twin engined single seat Navy fighter type designed with an eye towards carrier operation and with a tricycle landing gear, the Grumman Tigercat prototype, two of which were built, were first designed with Wright R-2600 Cyclones of 1600 horsepower but later were changed to take Pratt and Whitney R-2800 Double Wasps of 2100 horsepower each giving this ten ton fighter a high speed of over 420 mph. Armament was heavy; four .50 caliber machine guns in the nose and four 20mm cannon in the wing roots. First flight took place in November of 1943; the Navy Patuxent River NATC photo was taken in late December. The initial prototype crashed in May, 1944.

GRUMMAN F7F-1 (USN) A single place Tigercat planned as an F7F-1N night fighter with a nose radar, thirty four of which were built between April and October of 1944. These aircraft were part of a contract for 500 Marine Corps Tigercats ordered from Grumman; one of these -1s was revised to a -2N night fighter prototype. A few early planes had big propeller spinners like the XF7F-1. Normal gross weight was 21380 pounds; speed was similar to the X-model. An F7F-1 is shown in a Navy photo of July, 1945.

GRUMMAN F7F-2D (USN) The F7F-2 Tigercat was ordered in a -2N two place night fighter version to the tune of 66 aircraft between October, 1944 and August, 1945. The aircraft was convertable back to single seat. A few -2 airplanes were revised to a -2D drone or missile control variant as shown in the NAS Norfolk photo of August, 1950. Modifications included revised electronics and an aft cockpit using an F8F-1 Bearcat canopy. Missiles are shown underwing.

GRUMMAN F7F-2N (MFR.) A company photo of the F7F-2N night fighter shows the flush Plexiglas enclosure for a second crewmember. The -2N was first flown in October, 1944 as a conversion from a -1 airplane. The aircraft had the standard four .50 caliber and four 20mm gun armament and could carry two 2000 pound bombs or a one 2150 pound torpedo, or an assortment of rockets.

GRUMMAN F7F-3 (MFR.) The -3 version of the Tigercat was a single seat day fighter for the Marines. An additional fuel tank was used in place of a second crewman, the nose radar was omitted, and the vertical tail was revised. Two R-2800 Double Wasp engines of 2100 horsepower each provided the 11 ton normal gross weight fighter with a high speed of 435 mph at medium altitude. The photo illustrates the sleek clean lines of the big aircraft, 190 of which were delivered between March, 1945 and June, 1946. The Tigercat just missed World War II but some saw service later in Korea.

GRUMMAN F7F-3N (P. Bowers) Runup photo of a Marine F7F-3N night fighter, 60 of which were built, shows the large SCR-720 nose radar and the simple wing fold with outer panels carrying stand-offs for mounting rockets. The extra fuselage fuel tank was not included, but external drop tanks could be carried under the fuselage.

GRUMMAN F7F-3N (USN) Flight view of a US Marine F7F-3N night fighter shows the two place arrangement with radar operator aft under a flush enclosure. Normal gross weight was about 11 tons with a maximum overload of about 13 tons. The Marines used a few of these aircraft in a squadron in Korea in 1951-2. Fast (over 420 mph) and heavily armed (four 20mm cannon) the plane could be used for ground attack as well as night fighting.

GRUMMAN F7F-3P (P.Bowers) The photo shows a -3 airplane revised to an F7F-3P single seat photo reconnaissance model equipped with cameras up front in place of nose guns. Wingspan was 51 feet six inches and normal loaded weight 21900 pounds. High speed was close to 450 mph and service ceiling over 40000 feet. Normally an APN type ADF antenna was mounted on the rear turtledeck in a faired "football".

GRUMMAN XF7F-4N (R. Besecker) The experimental prototype of the F7F-4N carrier-based night fighter version is shown, the final Tigercat equipped with a carrier hook. The photo appears to be a "boneyard" shot. Although specifications called for Hamilton propellers of 13 feet one inch diameter, the decals on the blades appear to identify them as Aeroproducts types. Both engines and propellers rotated the same right hand looking from the cockpit.

GRUMMAN F7F-4N (R.Besecker) One of the final dozen production F7F-4N two place carrier-compatible Tigercat night fighters produced from September to November, 1946, making a total of 364 Tigercats produced. Changes from the -3N version included beefup of the inner wing panels and also of the landing gear for carrier operations. The gear had a longer extension and a shrink strut was used in effecting retraction.

GRUMMAN XF8F-1 (H.Andrews)
A Navy contract for two prototypes of a new lightweight single engine high performance fighter, the XF8F-1, was given to Grumman in late 1943. Use was to be made of a late model 2100 horsepower C version of the Pratt and Whitney Double Wasp engine with water injection capability. The photo shows a flight test view of the new Bearcat flown first on August 31, 1944 by Robert Hall.

GRUMMAN F8F-1 (R.Besecker)
Very large orders for F8F-1 Bearcats were gained by Grumman in October, 1944, and the first production aircraft was flown the last day of that year and delivered the same day. Though Bearcats were to be produced as F3M-1s by Eastern, contracts were cancelled at war's end. By that time 150 Bearcats had been produced but they never got to the war. A total of 659 F8F-1s were delivered, the last on August 29, 1947. The 2100 horsepower Double Wasp gave the 9400 pound normal gross weight Bearcat a high speed of about 425 mph. Early -1 aircraft had "breakaway" wingtips, later discontinued.

GRUMMAN F8F-1 (H.Andrews)
The F8F-1 Bearcat was armed with four .50 caliber wing guns. Two 1000 pound bombs or drop tanks could be carried or alternately four rockets. The Double Wasp engine drove a four blade Aeroproducts 12 foot seven inch diameter constant speed hydraulic propeller. Production aircraft had a dorsal fin added. First Bearcats went aboard carrier Langley just before the end of World War II.

GRUMMAN F8F-1B (D. Menard) The F8F-1B was a Bearcat equipped with four 20mm cannon (though the photo does not show these installed). A total of 224 -1B versions were produced by Grumman after a first flight in February, 1946. Production ran until January, 1948. The Bearcat was famous for its fast early climb capability, and at one time could beat anything up to 10000 feet.

GRUMMAN F8F-1N (USN) A night fighter Bearcat was put out only in small numbers; 13 aircraft were produced from May to November of 1946. As shown in the Navy photo the radome of the APS-19 radar system was hung from a bomb rack on the right wing. The -1N version had the standard four .50 caliber guns. A GR-1 autopilot was also installed.

GRUMMAN F8F-2 (MFR.) The new F8F-2 Bearcat had several changes from the -1. A new Double Wasp model delivered 2250 horsepower giving the 10500 pound fighter a high speed of just under 450 mph at altitude. The powerplant had a new variable speed supercharger and automatic engine controls. Standard armament was four 20mm cannon. The vertical tail height was increased by a foot. The -2 was first flown in June, 1947 and 283 fighters were delivered between October, 1947 and April, 1949, not including 69 F8F-2P photo reconnaissance versions and 13 F8F-2N night fighters. The Bearcat saw no Korean action, but some were provided to the French in Indochina.

GRUMMAN XF9F-2 (H.Martin) Grumman's first venture into the Navy jet fighter business was the straight wing Panther with two XF9F-2 prototypes. The XF9F-1 was an abortive tandem two place four jet engine design. The XF9F-2s used a single 5700 pound thrust British Rolls Royce Nene centrifugal flow jet engine later to be produced in the US by Pratt and Whitney. Gross weight was 5.5 tons and high speed 590 mph at sea level. First flown on November 24, 1947, one of the prototypes crashed, but not before successful testing was to lead to F9F-2 production.

GRUMMAN F9F-2 (R.Besecker) Photo of the F9F-2 Panther shows one of the wing root air intakes for the 5700 pound thrust Pratt and Whitney J42 engine, an Americanized Nene, a lowered perforated fuselage dive brake forward, the tail hook lowered, and one of the fixed wingtip external fuel tanks. First flown in February, 1949, production ran from then until the summer of 1951. The short landing gear elements allowed by jet propulsion instead of propellers are evident.

GRUMMAN F9F-2 (Author) A civil F9F-2 with rockets underwing photographed at NAS Willow Grove Pa. in October of 1988, the property of a well-heeled hobbyist. A total of 567 F9F-2 Panther variants were produced including some F9F-2D LABS fighter bombers and F9F-2P unarmed photo reconnaissance versions. A large number of F9F-2 jets were used by the Navy in the Korean War.

GRUMMAN F9F-3 (MFR.) Company photo of the F9F-3 Panther shows how wings folded. The -3 used an Allison J33 engine of 4600 pounds jet thrust planned as a backup powerplant in case the Pratt and Whitney J42/Nene development did not work out. Flown first in January, 1949, 54 aircraft were produced between December, 1948 and November, 1949. These aircraft were later revised to -2 models.

GRUMMAN F9F-3 (P.Bowers) Flight view of a test F9F-3 with two small bombs under the port wing and a nose boom shows clean lines of the aircraft. Performance of the -3 was somewhat less than the -2 because of less thrust from the Allison J33 engine. Though wingtip tanks could not be dropped a high speed fuel dump system for these tanks was installed.

GRUMMAN F9F-4 (R.Stuckey) The next version of the Panther was the F9F-4 originally powered by an Allison J33 jet engine of 6250 pounds thrust and lengthened by about a foot and a half to get more internal fuel space. The vertical tail height was increased by ten inches. First flight of the F9F-4 took place in August of 1950, and 109 of this model were delivered between September of 1951 and June of 1952. All F9F-4s were later converted to -5 models. Normal gross weight of the F9F-4 was 17700 pounds; high speed was about 590 mph.

GRUMMAN F9F-5 (USN) The Navy photo shows a flight of F9F-5 Panther jets from the carrier USS Boxer, CVA-21, in the Western Pacific on August 16, 1955. The F9F-5 was the last straight wing Grumman jet fighter and was a real workhorse in the Korean conflict. After a first flight in December, 1949 a total of 616 F9F-5 jets were delivered between November, 1950 and January, 1953. The -5 was similar to the -4 except for use of the 7000 pound thrust Pratt and Whitney J48 giving a maximum speed of about 600 mph. There were also three dozen F9F-5P photo reconnaissance jets produced in 1951-2.

GRUMMAN F9F-5KD (R. Besecker) As the straight wing Panthers became obsolete some were converted to drone aircraft or drone controller aircraft or both. The photo shows an F9F-5KD apparently convertable to either drone or controller at NAS North Island, Ca. in January, 1961.

GRUMMAN XF9F-6 (MFR.) The first operational swept wing jet fighter for the Navy, the third of three prototypes of which is shown, was the XF9F-6. These were F9F-5 conversions with new swept outer wing panels and tail. The Navy needed a swept wing jet fighter badly after encountering swept wing MIG-15s in Korea and ordered the conversion early in 1951. First prototype flight was in September of that year.

GRUMMAN F9F-6/F-9E (J.Davis) An F9F-6 Cougar in flight test with a boom on the nose and bombs on the wing store stations. Four 20mm cannon again comprised the gun armament. The 6250 pound thrust Pratt and Whitney J48 jet engine gave the nine to ten ton Cougar a maximum speed of 650 mph down low. and 600 mph at altitude. A total of 747 -6 Cougars were delivered in 1953 and 1954. Except for a very few aircraft the Cougar was too late to serve in the Korean fighting.

GRUMMAN F9F-6P (P.Bowers) A Marine photo reconnaissance version of the -6 Cougar is shown, one of 60 aircraft with cameras located in the nose. First flown in January, 1953, the 60 planes were delivered between June of 1954 and March of 1955. A flat glass panel enclosing the camera bay can be seen at the nose of the aircraft.

GRUMMAN F9F-7/F-9H (NACA) An F9F-7 Cougar on test at an NACA facility is shown with a boom on the nose. A total of 168 of this version were delivered between April, 1953 and June of the next year after a first flight in March of 1953. The aircraft was essentially a -6 with an Allison J33 jet engine of 6350 pounds thrust. This was a backup engine for the J48 of the F9F-6.

GRUMMAN F9F-8/AF-9J (R.Besecker) The final Cougar model was the F9F-8, essentially an F9F-6 with a new wing incorporating a cambered leading edge, increased fuselage length to carry more fuel with an aerial refueling probe in the nose, and an all-flying horizontal tail. Initial flight of the -8 was in January, 1954 with production of 601 fighters taking place from early 1954 to early 1957. Using the same Pratt and Whitney J48 of 6250 pounds thrust, the -8 Cougar had essentially the same performance as the F9F-6.

GRUMMAN F9F-8B/AF-9J (R. Besecker) Shown at NAS Glenview, Il. in April of 1961 is an F9F-8B fighter-bomber version equipped with the new low altitude (loft) bombing system (LABS). Two YAF-9J aircraft were converted initially from F9F-8s, and later larger numbers were changed over. The aircraft were modified in the field with service changes.

GRUMMAN F9F-8P/RF-9J A Marine photo reconnaissance version of the F9F-8 Cougar is shown getting a washdown and shows two of the several camera windows in a photo nose. Seven cameras were located to take pictures from forward, below, and to the sides. The prototype was first flown in February of 1955 and production of 110 aircraft lasted from August, 1955 to July, 1957. Using the same J48 engine high speed was 637 mph at 5000 feet.

GRUMMAN F9F-8T/TF-9J (R.Besecker) A two place trainer version of the F9F-8 Cougar used by Navy Attack Squadron VA-126 in April of 1965. The new tandem cockpit arrangement is readily apparent; the trainee seat was forward. Only two 20mm cannon were installed instead of the more standard four, but bombs or rockets could be carried underwing and aerial refueling capability was included to make this version fully combat operational if required.

GRUMMAN F9F-8T/TF-9J (USN) Close-up view of the two place trainer shows many aircraft details. Note the long canopy slides all the way aft from over both cockpits. It is not clear the second man knows where to step! A wing fence is shown on the port wing to prevent spanwise airflow. A total of 400 trainers were built from February, 1956 to February, 1960. High speed was 630 mph at 25000 feet and service ceiling was 43000 feet.

GRUMMAN XF9F-9 (H. Andrews) Although a completely new fighter design the three Tiger prototypes were originally kept in the F9F- series; later this error, perhaps deliberate, was corrected and the new aircraft turned into F11F-1s. First flown on July 30, 1954, the initial delivery to the Navy was made November 15, 1954. The aircraft had the area rule applied in its design for efficient transonic flight. Power was supplied by a 7800 pound thrust Wright J65 engine.

GRUMMAN XF9F-9 (H.Andrews) A Navy flight view of an XF9F-9 prototype shows some of the lines of the later F11F-1 Tiger. Note the drag brake extended under the belly to slow down for the photo plane. A Cougar is flying chase. Delivered first in November, 1954, the airplane was one of three experimental prototypes for the Tiger.

GRUMMAN XF10F-1 (MFR.) Ordered back in April of 1948, the Grumman XF10F-1 Jaguar went through a series of configuration changes and was the first attempt at an operational Navy jet fighter with variable sweep wings. That was a way to cover both low and high speed flight regimes . Along with a servo-controlled variable incidence tee-tail with a delta plan horizontal surface, the XF10F-1 used wing spoilers with small feeler ailerons for roll control. The pudgy jet was to use an afterburning Westinghouse J40 engine of 10900 pounds thrust giving an estimated high speed of about 710 mph. Actually a lower thrust J40 was used.

GRUMMAN XF10F-1 (MFR.) First flown in May of 1952, there were persistant problems in flight test because of incorporation of so many novel design features. Wingspan varied from 50 feet seven inches down to 36 feet eight inches with fully swept wings. Gross weight was just over 14 tons; armament was four 20mm cannon. At one time 140 aircraft were on order. As it happened only one prototype flew; another was 90 percent complete and a third 60 percent complete when the program was cancelled in mid-1953.

GRUMMAN F11F-1/F-11A (P.Bowers) The production version of the XF9F-9 prototypes, the F11F-1 Tiger flew first on July 10,1954 and 199 of the fighters were produced from November, 1954 until January, 1959. In 1962 the Tigers became F-11As. Powered with a Wright J65 jet engine of 10500 pounds thrust in afterburner, the Tiger had a high speed of about 750 mph and a maximum gross weight of about 11 tons. The jet's first line service time was short; it served a few squadrons from 1959 to 1961.

GRUMMAN F11F-1/F-11A (R.Sullivan) The F11F-1 Tiger was at least as famous for its use by the Navy's Blue Angel Flight Demonstration Team as for its squadron operational use. The photo shows aircraft #5 of the Angels at MCAS Quantico, Va. in September of 1976. A famous Tiger incident was when a pilot was shot down by overtaking his own bullets in a dive. The aircraft ended up in a crash landing.

GRUMMAN F11F-1F (USN) Two airplanes were completed as F11F-1F fighters using the General Electric J79 engine of 17000 pounds thrust giving a big increase in maximum speed to well over 1300 mph. Zoom climbs were made to almost 80000 feet. First flown in May, 1956, other changes from the production F11F-1 were a larger fuselage and longer extended engine air intake ducts. With new fighters coming along there was no further Tiger production.

GRUMMAN F-14A (R.Dean) The F-14A Tomcat won a new fighter competition after the failure of the F-111B program. A two place twin turbofan swing wing air combat fighter, the Tomcat has been the Navy's front line equipment since full fleet introduction in 1974. Equipped with a Hughes AWG-9 Airborne Weapons Control System and up to six Phoenix missiles, the Tomcat can attack several targets at the same time at long range. An early Tomcat is shown at a mid-1970s airshow.

GRUMMAN F-14A (Author) One of many F-14As parked at NAS Norfolk, Va. on February 23, 1976. The story was that engine problems had caused a Tomcat stand-down. First flown on December 21, 1970, the initial aircraft crashed on its second flight. Flight tests resumed in May, 1971. A total of 552 F-14A Tomcats were produced between September, 1972 and March, 1987.

GRUMMAN F-14A (MFR.) The photo shows three early F-14A Tomcats demonstrating the variation possible in wing sweep angle from 20 degrees to 68 degrees. The wing could be swept further to 75 degrees to minimize carrier stowage width in the hangar. Twin vertical tails provided the required tail area without folding for stowage.

GRUMMAN F-14A (USN) An F-14A Tomcat with wings ready to be extended from the stowed swept position. The fighter is from Squadron VF-2 readying for steam catapult launching from the carrier USS Enterprise, CVAN-65 on April 28, 1975. The F-14A mission was to provide cover and support for the evacuation of Saigon, RVN, at the end of the Vietnam War.

GRUMMAN F-14A (USN) Another F-14A of VF-2 is shown lauching from CVAN-65 on April 28, 1975 while operating in the South China Sea in support of the Saigon evacuation. Wings are extended in the slow speed configuration and full span flaps are lowered. Slab horizontal tails (stabilators) are positioned ready for climbout. Spoilers just ahead of the flaps provide low speed roll control.

GRUMMAN F-14A (USN) An F-14A of Navy Squadron VF-1 flies over the coast near San Diego, Ca. in January of 1974. Two other Tomcat versions have been produced. Thirty-eight F-14A "plus" models came out from November, 1987 to February, 1990 powered by General Electric F110 turbofans. In May, 1991 these were redesignated as F-14Bs. Then from March, 1990 to July, 1992 a total of 127 new F-14D Tomcats with two 27000 pound thrust General Electric F110 turbofans giving speed performance around Mach 2.0 were produced with upgraded avionics.

GRUMMAN F-14A (USN) A Soviet TU-95 Bear D long range reconnaissance aircraft is escorted by an F-14 Tomcat during exercise Teamwork 80. The Tomcat is from Fighter Squadron VF-84 embarked on USS Nimitz, CVN-68. These interceptions were quite common during the Cold War.

EBERHARDT FG-1 (USN) A Navy photo of July, 1927 shows the landplane version of an Eberhardt FG-1 Comanche fighter prototype with upper wing sweepback and lower wing swept forward. Provided for test to the Navy, the Comanche was evaluated and many changes were made but no production resulted. The FG-1 could be converted to a single float seaplane. Powered by a Pratt and Whitney 425 horsepower Wasp engine, the approximately 3000 pound Comanche had a landplane high speed of about 150 mph.

GOODYEAR FG-1 (MFR) The equivalent from Goodyear of the Vought F4U-1A Corsair fighter is shown in flight in 1943. Goodyear began supplying FG-1s in April of that year, part of the Vought/Goodyear/Brewster consortium for Corsair production. Powered by the Pratt and Whitney Double Wasp engine of 2000 horsepower, the six to seven ton gross weight FG-1 had a maximum speed of over 410 mph at a medium altitude.

GOODYEAR FG-1D (R.Stuckey) The Goodyear-produced Corsair equivalent to Vought's F4U-1D was the FG-1D with hundreds produced. The D version had an uprated 2250 horsepower Double Wasp giving it a maximum speed over 420 mph and incorporating twin inboard wing pylons for bombs or external fuel tanks. The photo shows an FG-1D post-war from the Naval Reserve at Mineapolis, Mn. with rocket stand-offs underwing.

GOODYEAR FG-1D (Author) A well restored civil FG-1D Corsair is shown during an airshow at NAS Willow Grove, Pa. in October of 1988. Goodyear turned out to be an excellent co-producer of Corsair fighters during World War II as opposed to Brewster.

GOODYEAR XF2G-1 (MFR.) Goodyear received an early 1944 contract for over 400 F2G- Corsairs with Pratt and Whitney R-4360 Wasp Major four row 28 cylinder engines of 3000 horsepower and a bubble canopy. The contract was cut to ten four-gun airplanes, five XF2G-1 land-based models as shown in the photo and five carrier-equipped XF2G-2 versions put out in 1945. With a speed of 400 mph or more over the whole altitude spectrum, these aircraft were fast. They were later sold as post-war surplus and some found fame as air racers.

HALL XFH-1 (USN) Hall Aluminum produced an experimental Pratt and Whitney Wasp powered fighter delivered to the Navy for evaluation in mid-1929. Its most noteable features were an all metal monocoque aluminum alloy fuselage sealed for water-tightness and a jettisonable landing gear. If nothing else the plane was designed for emergency water landings! A 2500 pound airplane loaded with a high speed of about 150 mph, the XFH-1 had an engine failure and crashed (in the water and it floated!) in early 1930, ending the naval career of the XFH-1.

MCDONNELL XFH-1 (A.Schmidt) The FH-1 Phantom from McDonnell was the first operational carrier jet fighter starting with a mid-1943 Navy contract. A single place twin jet aircraft, it was originally designated the XFD-1 but soon changed to FH-1 to avoid conflicts with Douglas. After testing 60 production aircraft were ordered with the first delivered in July, 1946. Using two Westinghouse J30 jet engines the Phantom served a few operational squadrons until about 1950.

MCDONNELL FH-1 Shown in flight the FH-1 Phantom displays the buried wing root installation of the Westinghouse jet engines and the fairings on the nose over the four .50 caliber machine guns. A large conformal drop tank could be attached under the fuselage. Powered by two J30 jets the five to six ton gross weight Phantom had a maximum speed of approximately 500 mph.

MCDONNELL XF2H-1 (MFR.) When this photo was taken the new Mc Donnell fighter, ordered in March of 1945, was labeled an XF2D-1 but was shortly changed to XF2H-1 so as not to interfere with Douglas designations. An early 1947 picture shows test pilot Bob Edholm alongside one of two Banshee prototype twin jet single place fighters. The aircraft was essentially a larger more powerful Phantom.

MCDONNELL XF2H-1 (MFR.) With the XF2D-1 designation still printed on the rudder, the new Banshee with two Westinghouse J34 jet engines buried in the fattened straight wing roots and dihedral used in the horizontal tail to avoid jet exhaust, shows off clean lines. First Banshee flight was on January 11, 1947. Part throttle operation of the jet engines yielded high specific fuel consumption; it was more economical to operate at a higher power on a single engine, but pilots could be reluctant to shut one down.

MCDONNELL F2H-1 (R. Besecker) Successful testing of the prototype aircraft led to an April, 1947 Navy order for 56 production F2H-1 Banshee fighters which eliminated the horizontal tail dihedral as unnecessary as can be seen in the photo. Thirty aircraft were procured with Fiscal Year 1947 funds and the remainder with FY 1948 money. Powered by two Westinghouse J34 jets each of 3000 pounds thrust, the seven to nine ton gross weight F2H-1 had a maximum speed of over 580 mph. Armament was four 20 mm cannon located in the nose.

MCDONNELL F2H-2 (P. Bowers) An F2H-2 version of the Banshee in the Reserves is shown in the photo which emphasizes the short landing gear allowed by jet installation having no propeller to clear. Changes in the F2H-2 involved J34 engines uprated to 3250 pounds of thrust each, a longer fuselage to house additional fuel, provision for large wingtip mounted fuel drop tanks, and two external store stations. Gross weight went up to nine to eleven tons; high speed performance was similar to the -1. A total of 334 F2H-2 Banshees were delivered from late 1949 to early 1952. One hundred and eight were procured with FY 1949 and 146 with FY 1950 funds. Many were active in the Korean War starting in 1951.

MCDONNELL F2H-2N (MFR.) Test pilot Bob Edholm sits in the cockpit of a new F2H-2N night fighter version of the Banshee with the substantially sized wing flaps lowered. Landing speed was about 100 mph. Fourteen F2H-2N aircraft were produced using FY 1949 funds. The aircraft nose was modified to reposition the 20mm cannon and allow installation of APS-6 radar components up front under the radome.

MCDONNELL F2H-2P (P.Bowers) Another variant of the -2 Banshee ("Banjo") was the F2H-2P photo reconnaissance version with 88 aircraft delivered, 58 produced with FY 1950 funding and the remainder with the next year funds. The ports for three of the six side-looking cameras can be seen in the photo of this long nose version. No guns were carried on the F2H-2P airplanes. Note that wingtip fuel tanks are mounted.

MCDONNELL F2H-3 (P.Bowers) Shown at a 1956 airshow is an F2H-3 Banshee fighter of Navy Squadron VF-64 with wings folded and wingtip fuel tanks up in the air. An all-weather version of the Banshee, the F2H-3 retained the four 20mm cannon just aft of the APQ-41 radar nose; the gun ports can be seen. Another extension of the fuselage provided even more space for internal fuel and more wing store stations were added. The aircraft had a normal gross weight of 10.5 tons and a high speed of 580 mph. A total of 250 -3 versions were procured in 1952.

MCDONNELL F2H-4 (USN) A Navy photo of October, 1955 shows two F2H-4 Banshees, one equipped with wingtip fuel tanks and the other without. Both aircraft carry nose probes for aerial refueling capability as did most earlier versions in a retrofit. McDonnell built 150 of the -4 models with the last delivered in October, 1953.

MCDONNELL F2H-4/F-2D (G. Williams) Subsequent to their withdrawal from front line service, the remaining F2H- aircraft were redesignated as F-2s in 1962. The F2H-4 was then an F-2D. Shown in a September, 1955 photo is the final model of the "Banjo", the F2H-4 showing details of the updated radar nose and cannon installation. The additional wing store stations are shown. This -4 model has apparently not been fitted with an aerial refueling probe. The Westinghouse J34 engines were uprated to 3600 pounds of thrust each. Maximum weight increased to 29000 pounds; maximum speed was 610 mph.

MCDONNELL XF3H-1 (MFR.)
The new McDonnell Navy fighter of 1951 was the XF3H-1 Demon interceptor shown in this company photo of July 6, 1951, just a month before first flight as one of two prototypes ordered in late 1949. With a needle nose and swept wings and tail surfaces the new Demon looked a winner and plans were made for high production. The intended version of the Westinghouse J40 engine did not materialize to the detriment of the program.

MCDONNELL F3H-1N (USN) The first production version of the Demon was a night fighter which turned into a procurement and flight test disaster. The first flight took place in January, 1954. The photo shows an F3H-1N Demon at NATC Patuxent river, Md. on September 17, 1954. Lack of thrust from the J40 engine was a principal cause for many crashes during testing with loss of life, and in 1955 the Demon was grounded. Of the aircraft built 21 acted as ground trainers and 29 were revised to use an Allison J71 engine.

MCDONNELL F3H-2/F-3B (USN) The developed model of the Demon was the F3H-2, this time with a revised wing with over 70 square feet greater area and powered by an Allison J71 jet engine of 14400 pounds thrust, a very significant increase over the F3H-1. Deliveries started in 1957 and ran through 1959, and Demons remained in frontline service through the early 1960s. Equipped with both four 20mm guns in the fuselage and underwing missiles the 39000 pound gross weight -2 Demon had a high speed of about 700 mph at sea level and 630 mph at altitude.

MCDONNELL F3H-2M/MF-3B (P.Bowers) The first missile-carrying version of the Demon, 80 -2M versions were delivered in the mid-1950s. An aircraft from the Naval Missile Test Facility at Point Mugu, Ca. is shown with four Sparrow missiles carried on underwing pylons. The ports for two of the 20mm cannon are covered in the area just under the engine intake. The flat area at the rear of the fuselage shows the range of angular travel of the all-flying horizontal tail.

MCDONNELL F3H-2N/F-3C (P.Bowers) An all-weather version of the Demon was the F3H-2N which started operations in 1956. A total of 142 aircraft were built. The photo shows several features of this four-cannon fighter of VF-142 including an open air brake alongside the fuselage, a variable area exit nozzle for the Allison J71 engine, and the all-flying tail in the full nose-up position. McDonnell ended up providing a total of well over 500 Demon fighters to the Navy.

MCDONNELL F4H-1 (USN) One of the 23 early F4H-1 Phantom II test aircraft is shown. Originally slated to be the AH-1 attack aircraft with an order for two single place prototypes in October of 1954, the designation was changed in May of the next year to F4H-1 with a change of mission to a long range missile-equipped fighter. The photo was taken in January, 1960. First flight of the two place twin engine F4H-1 was in May of 1958.

MCDONNELL F4H-1 (H.Andrews) An early development Phantom II (note the name on the fuselage) shown during carrier compatibility testing. The Phantom II won out in competition with the Vought F8U-3 fighter. Wing leading edge and trailing edge flaps are shown deflected. Basic wing sweep was 45 degrees; outer panels has a 12 degree dihedral. The all-moveable horizontal tail was set at 23 degrees of anhedral. The F4H-1 later turned into an F-4A and reached a speed during tests of Mach 2.60.

MCDONNELL F4H-1F/F-4A (R.Besecker) An F4H-1 is shown at NATC Patuxent River, Md. This Navy and Marine all-weather fighter was powered by two General Electric J79 jet engines each of 17000 pounds thrust with afterburners operating, and carried four semi-submerged Sparrow III missiles. It used a Westinghouse APQ-72 air interception radar in the nose. A total of 696 F-4A and F-4B Phantoms were delivered. Normal gross weight was about 22 tons; high speed was just under 1500 mph at high altitude and well over 800 mph at sea level. The early F4H-1Fs were redesignated F-4As in 1962.

MCDONNELL F-4B (AAHS) The F-4B Phantom, the initial operational production model first flown in March, 1961, is represented by an aircraft of Marine All-Weather Squadron VMF(AW)-314 in flight. A total of well over 600 F-4Bs were delivered from mid-1961 through 1966. Normal gross weight of an F-4B was 22 tons while maximum weight was almost 28 tons. Besides Sparrow or Sidewinder missiles the Phantom could accomodate a nuclear store or up to six tons of iron bombs.

MCDONNELL F-4B (R.Besecker) One of the first batch of production F-4B Phantoms is shown with wingtips folded at PAX River in April of 1974, about twelve years after initial deliveries. The hatch arrangement of the tandem cockpits is shown. The pilot was up front and the radar intercept officer sat in the rear. Dual controls were not normally employed. The aircraft could accelerate from Mach 1.0 to Mach 2.0 in about 3.5 minutes. By 1964 the F-4B had equipped 20 squadrons.

MCDONNELL F-4G (R. Besecker) An F-4G test version of the Navy Phantom is shown at a display in August of 1968, probably at NATC PAX River, Md. The 12 G versions were initially F-4Bs in a development incorporating an ASW-21 data link system and served in Vietnam in early 1966 aboard the USS Kitty Hawk. Later over 100 USAF F-4E types were converted to F-4G Wild Weasel aircraft.

MCDONNELL F-4J (J.Weathers) An F-4J Phantom of Navy Squadron VF-21 from the carrier USS Ranger is shown at NAS New Orleans in April, 1977. This model was a development of the F-4B with both interception and ground attack capabilities. New droopable ailerons and a slotted tail provided a reduced speed of approach. The Phantoms used a Westinghouse AWG-10 fire control system and a new bombing system. The G model first appeared in May of 1966. This version was used by the Navy Blue Angels. The J was powered by two General Electric J79 engines each of 17900 pounds thrust giving the 24 ton fighter a high speed of about 1400 mph at altitude.

MCDONNELL F-4J (J.Weathers) Another view of the F-4J of VF-21 from the USS Ranger with a large centerline store, probably a fuel tank. The F-4J could climb to well over 40000 feet in a minute and had a normal ceiling of about 55000 feet. Armament could be six Sparrow III or four Sparrow III and four Sidewinder missiles in semi-submerged mounts or alternate loads on one fuselage and four wing store stations for bombs, mines, pods, or tanks. Many F-4Js were updated to F-4S versions with new electronics systems.

MCDONNELL F-4N (R.Dorr) One of over 200 updated F-4Bs or F-4Gs first flown in June, 1972 and the first delivered February 21, 1973. The F-4N of Navy Squadron VF-201 shown in the photo is an updated F-4G at NAS Mirimar in December, 1976. The N models had new electronics installed. The photo shows a good view of the retracted tail hook between the exhausts of the two engines, and also the wing fold.

MCDONNEL DOUGLAS FA-18A (USN) Originally based on the Northrop YF-17 design, the FA-18A Hornet was at the start a much revised Navy Air Combat F-18A fighter with first flight in November of 1978. Carrier qualification tests took place in late 1979. The Navy required a new fighter-attack plane besides the costly F-14A Tomcat to replace the F-4 Phantoms, and also to replace the Vought A-7 Corsair II light attack plane. The new aircraft finally became the FA-18A single seat twin turbofan engined multi-mission fighter. Eleven prototypes were built.

MCDONNELL DOUGLAS FA-18A (B. Holder) A current standard Navy combat aircraft which entered service in January, 1983, the FA-18A Hornet uses two General Electric F404 augmented turbofans each of 16000 pounds of thrust providing the 19 to 25 ton gross weight aircraft with a high speed of almost 1200 mph in clean condition at altitude. Armament consists of a 20mm M61 high rate of fire cannon along with a wide range of external stores on nine weapon stations. A total of 371 US Navy FA-18A and 534 FA-18C and two place FA-18D aircraft have been built and well over 1300 of these types have been exported.

MCDONNELL DOUGLAS FA-18B (S.Orr) Navy planning called for two two seat Hornets for every eleven single seat FA-18s so 39 FA-18Bs were built. The two tandem seat versions have six percent less internal fuel than the single place aircraft. The tandem seat arrangement of the FA-18B is well shown in this June, 1991 airshow picture where the single store is a drop tank on centerline. The port for the M61 gun is high on the nose just aft of the Hughes APG-65 radar. A retractable aerial refueling probe compartment is located forward on the right side.

MCDONNELL DOUGLAS FA-18B (J.Weathers) A two place Hornet of Marine Squadron VFA-125 is shown at NAS New Orleans on June 4, 1983. The photo illustrates the twin vertical tails, the twin engine variable exit nozzles for the F404 turbofan engines, and the slab horizontal tails called stabilators or tailerons, the latter name illustrating their use to assist in roll control. The wing is equipped with leading and trailing edge flaps as well as droopable ailerons to reduce approach speeds. The FA-18B is the two seat version of the FA-18A; the FA-18D is the two seat version of the FA-18C.

BOEING F/A-18E/F Super Hornet(MFR.) This version of the Navy Hornet is the nation's newest tactical strike fighter with the E version single seat and the F a two seater. A development contract was awarded in June of 1992, and ten test articles were fabricated including seven flying aircraft. Testing is now complete and low rate production has started. The type will enter fleet service in 2001. The Super Hornet is 25 percent larger than the Hornet and has substantially increased performance via use of two General Electric F414-GE-400 turbofan engines of 35 percent more thrust. Greatly increased fuel capacity provides for a 40 percent mission radius increase. Two additional weapons stations bring the total to eleven. A Ratheon APG-73 radar is in the nose. Boeing builds wings and forward fuselage, and does final assembly; Northrop Grumman builds the center and aft fuselage.

BERLINER JOYCE XFJ-2 (USN) The initial effort at a Navy fighter by Berliner Joyce and originally appearing as an XFJ-1 in the spring of 1930 after a May, 1929 Navy order for one prototype. A distinguishing feature was the strutted standoff of the lower wing from the fuselage coupled with the direct connection of the upper wing. Extensively damaged in a landing accident the aircraft was returned to the factory for repairs. It emerged again in modified form with cowl, spinner, and wheel pants as the XFJ-2 shown in the Navy photo of May 26, 1931, four days after new tests at Anacostia were started.

BERLINER JOYCE XFJ-2 (USN) Powered with a 500 horsepower Pratt and Whitney Wasp and with a gross weight of 2850 pounds, the XFJ-2 had a top speed of about 190 mph, over a 10 mph increase from the original model. The unusual lower wing arrangement may have provided a strong ground effect interference upon landing and contributed to difficulty in landing the plane. No more XFJ-2 fighters were procured.

BERLINER JOYCE XF2J-1 (USN) A single experimental two seat fighter ordered in June, 1930 but not delivered for almost two years, the XF2J-1 is shown in initial configuration in a Navy photo of January 26, 1933 with open cockpits and fixed landing gear. A 625 horsepower twin row Wright R-1510 engine powered the 4500 to 4800 pound gross weight fighter to a maximum speed of about 195 mph at 6000 feet. The plane was revised to incorporate enclosed cockpits, but could not compete with the retractable landing gear Grumman FF-1 already under order by the Navy and no more XF2J-1 aircraft were produced.

BERLINER JOYCE XF3J-1 (H.Thorell) The last Berliner Joyce fighter designed for Navy carrier use was the very attractive single place XF3J-1 biplane with wings of double eliptical planform, an enclosed cockpit, and fixed landing gear. A long chord cowl enclosed the 625 horsepower two row Wright R-1510 Twin Whirlwind engine in front of an all metal fuselage. The fighter finally appeared in early 1934 and was delivered that April, only a month before the Navy was to order a production lot of Grumman F2F-1 fighters with retractable landing gear.

BERLINER JOYCE XF3J-1 (USN) A Navy photo of May 23, 1934 shows the XF3J-1 ready for testing. A three bladed fixed pitch propeller is noteable. Vibration problems were encountered by the test plane which could achieve a high speed of 209 mph, over 20 mph slower than the contemporary Grumman F2F-1 in production. In addition the Wright R-1510 engine was a failure while the competitive Pratt and Whitney R-1535 Twin Wasp Junior was a relative success. No further production of the Berliner Joyce aircraft was pursued, and B/J shortly sold out to North American.

NORTH AMERICAN XFJ-1 (MFR.) A manufacturer's photo shows one of three XFJ-1 Fury prototypes for North American's first jet aircraft as far as first flight (September 11, 1946 with Wally Lien piloting) is concerned. Work had been initiated on the XFJ-1 in December of 1944. The three prototypes were accepted by the Navy in 1947. A fat fuselage straight wing aircraft powered by a General Electric J35 axial flow jet engine of 3820 pounds of thrust, the prototype Fury testing led to a Navy request for a small batch of production fighters.

NORTH AMERICAN FJ-1 (P.Bowers) A closeup view of one of 30 production FJ-1 Furys shows nose inlet details; the photo was taken after the FJ-1s were sent to Naval Reserve squadrons after a little over a year of evaluation at sea starting in March, 1948. Only one operational test squadron, VF-5A (later VF-51) was formed to evaluate FJ-1 carrier compatibility. It was placed on five different carriers at various times. Powered by an Allison-built J35 engine with 4000 pounds of thrust giving the 15600 pound maximum gross weight FJ-1 a high speed of 547 mph at low altitude, the armament of the fighter was six .50 caliber guns in the forward fuselage.

NORTH AMERICAN FJ-1 (USN) One of the FJ-1 Fury fighters shown operated, as the legend on the aircraft nose proclaims, by the NAS Oakland, Ca. Reserves starting in late 1948, this use to familiarize new pilots with jet aircraft. Though they performed a useful service in that respect their small numbers and unusual features, including a kneeling nose wheel for carrier stowage, made them difficult to maintain. Changing engines, a difficult job, had to be done frquently because the operating life of early jet engine components was short.

NORTH AMERICAN XFJ-2 (MFR.) One of three XFJ-2 Fury swept wing prototypes ordered in March, 1951 when the Navy was suffering from lack of a suitable swept wing carrier fighter to take on the likes of the Russian MIG-15s. A navalized USAF F-86E Sabre, the XFJ-2 incorporated wing folding and other features for carrier compatibilty, including an extensible nose gear strut to increase takeoff angle of attack. Armament was changed to four 20mm cannon. Service testing started in 1952. Though carrier qualified the XFJ-2 was heavier than desired and production aircraft were assigned to Marine land-based units.

NORTH AMERICAN FJ-2 (R.Besecker) A well-used FJ-2 land based Marine Fury of VMF-312 is shown in the photo. The wing leading edge slat is open, a tail bumper is down, and the carrier tail hook is removed. Gun ports for two of the four 20mm cannon are shown alongside the nose. Store stations underwing support drop tanks with stabilizing tail fins.

NORTH AMERICAN FJ-2 (R.Esposito) Another view of a Marine FJ-2 Fury looking just like the USAF F-86 Sabre. Powered by a General Electric J47 jet engine of 6000 pounds thrust, the FJ-2 grossed out at about 16500 pounds and had a high speed of about 675 mph at sea level. Span was 37 feet one inch and length 37 feet seven inches. A total of 200 FJ-2s were produced from late 1952 until late 1954.

NORTH AMERICAN FJ-3 (R. Besecker) An FJ-3 shown after assignment to the reserves shows wing changes where the leading edge slats of the FJ-2 were replaced by extended leading edges like its Sabre counterpart and a small wing fence was added. Another major change was use of a 7700 pound thrust Wright J65 jet engine. The aircraft looks well used.

NORTH AMERICAN FJ-3 (MFR.) A manufacturer's flight view of an FJ-3 with drop tanks shows an aircraft before squadron assignment and before wing modifications and added pylons. The FJ-3 was carrier-compatible and was used by many fleet squadrons starting in late 1954 and going into the later 1950s. A total of 389 FJ-3s were procured. The J65 powered fighters achieved a top speed of about 680 mph at sea level, and could climb to a ceiling of just under 50000 feet.

NORTH AMERICAN FJ-3 (MFR.) A manufacturers publicity photo of late May, 1956 shows the then-current Blue Angel team members in formation trying out new production FJ-3s, though these were not used in Angel public demonstrations. The FJ-3s also carried significantly more internal fuel than the FJ-2 aircraft.

NORTH AMERICAN FJ-3M/MF-1C (P.BOWERS) A modification of 80 FJ-3s to missile carriers with two extra pylons marked one of the first installations of missiles aboard Navy fighters. Two Philco Sidewinders could be carried. The photo shows an FJ-3M Fury of Navy Squadron VF-24. The FJ-3 was the navalized equivalent of the USAF F-86F, and came into service in 1956.

NORTH AMERICAN FJ-4/F-1E (P. Bowers) The final basic version of the Naval Fury with a 7700 pound thrust J65 engine was a new design in most all respects. Contracted for in October, 1953, two prototypes flew a year later with four wing pylons for Sidewinders. The fuselage was thickened to include 50 % additional fuel capacity and new thinner wing and tail surfaces were utilized. A total of 150 production Fury aircraft were made in the Columbus, Ohio plant of North American (the wartime Curtiss plant) starting in 1955 with production complete in early 1957. The FJ-4 Fury was flown by some Navy and many Marine squadrons. Performance was generally similar to the FJ-3 except for radius capability. In 1962 the FJ-4s were redesignated as F-1Es.

NORTH AMERICAN FJ-4B/AF-1E (MFR.) A flight view of the final Fury modification shows the ground attack version of the FJ-4 with six store stations for carriage of up to 4000 pounds of bombs, missiles such as Bullpups, fuel tanks, or rocket pods. New systems such as LABS allowed carriage and delivery of tactical nuclear stores and an aerial refueling capability was added as well as new speed brakes.

FJ-4B/AF-1E (R.Esposito) An FJ-4B serving with the Navy and Marine Reserves at NAS Willow Grove, Pa. in the 1960s is shown in the photo. This final Fury model served until 1962 in operational squadrons and then went to the Reserves. A total of 222 aircraft were produced, most all for the Marines.

NORTH AMERICAN FJ-4B/ AF-1E (J.Weathers) Flight view of the FJ-4B Fury shows the aircraft carrying Bullpup air-to-ground missiles. As many as five missiles could be carried with the remaining underwing store station occupied by a missile guidance pod. Production of the Fury series ended in 1958 after over 1100 FJ- fighters had been delivered to the Navy and Marines.

NORTH AMERICAN FJ-4F (MFR.) This company photo of April 29, 1957 shows one of two FJ-4 aircraft modified for testing auxiliary rocket power with that engine mounted low and aft in the fuselage. The rocket installation was a scheme to provide a burst of power to such fighter prototypes as the Vought F8U-3, but the project was never completed.

BELL XFL-1 (MFR.) The XFL-1 Airabonita was a navalized version of the Army's Bell P-39 Airacobra, though it was a very different aircraft with a new larger wing, revised radiator system, and tail down landing gear. With some misgivings the Navy went ahead with this liquid cooled engine fighter project. First flight was in May, 1940. Aft center of gravity and stability problems along with engine cooling problems plagued the design, and several flight test emergencies were encountered. With the relative success and higher performance of the new Vought Corsair fighter the Navy lost interest in the Airabonita, and in May, 1941 it was deemed not satisfactory for service. The 1100 horsepower Allison V-1710 engine powered the three ton gross weight fighter to a high speed of 333 mph.

EASTERN FM-1 (J.Weathers) An FM-1 Wildcat fighter is shown cruising off the west coast. This Wildcat version was built by Eastern Aircraft Division of General Motors and was essentially the design equivalent of the Grumman F4F-4. The first FM-1 was flown September 1, 1942 after an April contract. The essential difference of the FM-1 was installation of four wing .50 caliber guns in place of six as in the F4F-4. A total of 1150 FM-1s were produced through 1943, some for the British as Wildcat Vs. The two stage supercharged Pratt and Whitney Twin Wasp engine powered the four ton fighter to a high speed of about 320 mph.

EASTERN FM-2 (R.Besecker) A post-war photo of an Eastern FM-2 Wildcat is shown at Chino, Ca. with rocket standoffs underwing and two machine gun ports showing, a Hamilton Hydromatic propeller, and that never to be forgotten narrow tread manually retractable landing gear. Eastern manufacture of the FM-2, based on Grumman XF4F-8 prototypes, started in the fall of 1943 and continued through to the end of World War II.

EASTERN FM-2 (UNK.) One of many post-war civil FM-2 Wildcats is shown in the photo (*above*). It was powered by a 1350 horsepower Wright nine cylinder R-1820 engine with a single stage two speed supercharger which usually drove a Curtiss Electric propeller as shown. The FM-2 was by far the most-produced Wildcat with almost 4800 built by Eastern. Many were used on small US and British escort carriers assigned to ASW duties.

EASTERN FM-2 (Author) A runup photo taken at Coatsville, Pa. in October 1974 of a civil FM-2 Wildcat owned by Mr. Alexi DuPont shows the typical contours of a Wildcat changed only by installation of modern avionics antennas. Navy FM-2 aircraft had a normal gross weight of about 7500 pounds and a high speed of about 330 mph. Service ceiling was approximately 35000 feet.

SEVERSKY NF-1 (A.LaClair) In September of 1937 the Navy tested a navalized version of the Army's P-35 pursuit plane which Seversky called an NF-1 for naval fighter one. No real Navy designation was applied because the airplane was not purchased. Powered by a Wright R-1820 Cyclone engine the NF-1 high speed of 267 mph and other performance items did not impress the Navy, and no Severskys were ordered.

RYAN XFR-1 (USN) During World War II the Navy became interested in the potential of composite powered fighters enjoying the advantages of both piston and the new jet engine power both for carrier compatibility (the pure jets did not have enough thrust to take off from a carrier without catapulting) and dash speed capability. Since Ryan was not producing combat aircraft for war use they received a February, 1943 contract for three XFR-1 Fireball fighter prototypes powered by a Wright Cyclone piston engine up front and a 1600 pound thrust General Electric jet engine in the tail. One of the three prototypes is shown on a carrier deck in a Navy photo dated May 12, 1945.

RYAN FR-1 (R.Besecker) Based on successful testing of the XFR-1 prototypes the December, 1943 order for 100 FR-1 Fireball production aircraft (at one time 1000 planes were on order) was implemented and a production FR-1 flew first in June of 1944. Among the changes were revised tail surfaces with the horizontal tail lowered to the fuselage. A new laminar flow airfoil wing along with overall flush riveting were features incorporated. As shown in the photo the FR-1 incorporated a tricycle landing gear. All these were new items for Navy fighters. The flush riveting was to give problems in service however, and structural failures occurred.

RYAN FR-1 (P.Bowers) Another view of the FR-1 Fireball shows the cleanly cowled Wright R-1820 Cyclone engine driving a Curtiss propeller with blade cuffs and one of the small air intakes for the jet engine at the wing root. A cut-back to 66 Fireballs was made, and all were produced in 1945. Armament was four .50 caliber wing guns. Bombs, tanks, or rockets could be carried underwing. The 1425 horsepower Cyclone alone gave the five ton gross weight plane a high speed of 295 mph; with the 1600 pounds of General Electric I16 jet engine thrust added the dash speed went up to just over 400 mph at medium altitude.

RYAN FR-1 (P.Bowers) This view of an FR-1 Fireball makes the aircraft look completely conventional. A special evaluation squadron, VF-66, was formed to use the Fireballs but was short-lived. Later the same squadron people were transferred to VF-41 which went aboard escort carriers, but the war ended before they could become fully operational. In addition problems arose which caused accidents and enthusiasm waned for the composite type of fighter. The Fireball was the first plane to land on a carrier solely on jet power when the piston engine started to fail.

RYAN FR-1 (MFR.) No story of a composite powered fighter would be complete without a picture showing it flying with the propeller stopped and feathered and such a photo of the FR-1 is shown. Such operation was not normal, however. The Cyclone was to be run continuously for cruising and the jet engine to be lighted up for bursts of speed as required. FR-1 operation stopped in 1947.

RYAN FR-4 (MFR.) In late 1944 one of the FR-1 Fireballs was considerably modified to test a more powerful jet engine in the tail, and this variant was labeled an FR-4. Major modifications were the use of dual flush air inlets forward on either side of the fuselage for the jet and inclusion of the more powerful Westinghouse J34 jet engine of about 3400 pounds thrust. The airplane was marginally faster but the piston engine started limiting performance.

RYAN XF2R-1 (MFR.) The solution to the XFR-4 piston engine limitations was installation of a General Electric XT31 turboprop engine of 1700 horsepower plus 500 pounds of jet thrust, the first such powerplant in a Navy airplane. The I16 tail jet engine was retained. The result was the XF2R-1 Dark Shark with a new nose section and a larger dorsal fin as shown in the photo. Development troubles with the early XT31 turboprop were sufficient to keep the Dark Shark from being a production item.

RYAN XF2R-1 (MFR.) Flight view of the XF2R-1 shows the new long nose of the prototype. The destabilizing effect of the long nose and new four bladed propeller required a larger dorsal fin as shown in the photo. When the powerplant was working properly the Dark Shark could reach a speed of nearly 500 mph. The T31 was not a successful turboprop engine, however, and never reached production on any airplane.

NORTHROP XFT-1 (MFR.) The Navy ordered a new single seat low wing all metal monoplane from Northrop in May of 1933 based on that manufacturors experience with all metal construction. The stubby little fighter, labeled an XFT-1, was in the modern mold except for Northrop fixed panted landing gear. First flown in January, 1934 it was delivered two months later. Tests of the 625 horsepower Wright R-1510 engine powered 3750 pound fighter showed a high speed of 235 mph, but revealed flying quality problems, including a propensity to spin easily. The photo shows the XFT-1 in January, 1934, two days after a first flight.

NORTHROP XFT-2 (MFR.) The little (32 foot span) Northrop monoplane took a long time in development. A major revision of the airplane was made at Navy behest to an XFT-2 with a new engine, the Pratt and Whitney R-1535 Twin Wasp Junior of 650 horsepower and a changed landing gear design. Flying qualities had not improved, and not long after the new version had been delivered to the Navy in March, 1936, the date of this photo, the fighter was grounded. The Northrop test pilot decided on his own to fly the plane back to the west coast Northrop plant, and in the process crashed it in Pennsylvania.

VOUGHT VE-7 (MFR.) A very good Vought design, the VE-7 was built by both the parent company and the Naval Aircraft Factory in 1920 to a total of over 125 aircraft, and several versions were used by the Navy. The variant used as a fighter with the front cockpit covered over and a single fixed forward .30 caliber machine gun was the VE-7S (or VE-7SF if flotation gear was added). The 180 horsepower Wright E-2 engine gave the 2100 pound VE-7S a maximum speed of about 115 mph.

VOUGHT FU-1 (Wright) A development of the Vought UO- two seat observation series of biplanes, itself an improvement of the VE-7 airplanes, the FU-1 single seat biplane, 20 of which were originally ordered by the Navy as UO-3s in mid-1926 and employed as two-gun fighters for about two years. Powered by a 220 horsepower Wright J-5 Whirlwind engine with a supercharger (see part of the ducting below the engine), the FU-1 had a high speed of about 120 mph.

VOUGHT XF2U-1 (MFR.) Vought's entry into a two seat fighter competition is shown in an engine runup photo. Delivered for tests in June, 1929 after a prolonged company development, the XF2U-1 was armed with two fixed forward firing .30 caliber guns in the upper wing and a single weapon for the rear gunner. Powered by the 450 horsepower Pratt and Whitney Wasp engine, the fighter had a gross weight of 4200 pounds and a maximum speed of 146 mph. Assigned to the Naval Aircraft Factory, the single aircraft was destroyed in an early 1931 crash.

VOUGHT XF3U-1 (MFR.) There was no lack of two seat Navy fighter competitors in the early 1930s. One of these was a Vought entrant based on a Navy design. The Vought XF3U-1 was ordered in June, 1930 and had a first flight in May of 1933. A clean cut aircraft with enclosed cockpits and fixed landing gear, the plane, shown in a September, 1933 company photo, was soon considered as a possible scout bomber rather than a fighter, and with modifications evolved into the Vought XSBU-1.

VOUGHT XF3U-1 (MFR.) A company flight photo of December, 1933 shows the configuration of the XF3U-1 with cockpit enclosures open. Twin .30 caliber fixed machine guns were mounted in the forward fuselage with a single gun in the rear. Powered by a Pratt and Whitney R-1535 Twin Wasp Junior engine the 4600 pound gross weight XF3U-1 had a top speed of 214 mph. The prototype ended up as an engine test bed airplane at the Pratt and Whitney plant, but testing of the prototype started the process of a large SBU-1 scout bomber order for Vought.

VOUGHT XF4U-1 (MFR.) The single prototype for a successful World War II Navy and Marine Corps fighter aircraft, the XF4U-1 design was based on use of the new 1850 horsepower Pratt and Whitney XR-2800 Double Wasp two row 18 cylinder engine with a two stage mechanical supercharger. Ordered in mid-1938, the single place low gull wing monoplane XF4U-1 Corsair had its first flight in late May of 1940. After a July forced landing accident the prototype was repaired and on October 1,1940 attained a speed of just over 400 mph. This performance led to orders for production Corsairs the next year.

VOUGHT F4U-1 (MFR.) A Vought photo of September 20, 1943 shows the initial production configuration of the F4U-1 Corsair. Major differences from the XF4U-1 involved moving most fuel to the fuselage from the wings and moving the cockpit aft to make room along with a six .50 caliber gun wing armament. The combination of the aft cockpit and the low "birdcage" canopy led to complaints about poor pilot visibility. The Pratt and Whitney Double Wasp provided the six to seven ton gross weight F4U-1 fighter with a maximum speed of about 415 mph at medium altitude.

VOUGHT F4U-1A (MFR.) In order to improve visibility Vought designed a new raised cockpit canopy and a lengthened tail gear along with other modifications-this after almost 1000 Corsairs had been produced up to the summer of 1943. The A designation was used but was not official. These changes were introduced somewhat after the Corsair's combat introduction in February, 1943 in the Southwest Pacific. Another problem on early Corsairs was a stiff landing gear causing planes to bounce upon landing. The Marines used the aircraft as a land-based fighter only through a large part of the war.

VOUGHT F4U-1D (MFR.) A water injected version of the R-2800 engine providing up to 2250 horsepower, provision for carrying rockets on late models, and two stores pylons under the wing center section, shown loaded with two 1000 pound bombs in the flight photo characterized the most widely used F4U-1D version (1685 aircraft) of the Corsair. This variant was produced from early 1944 to early 1945, and could get to a speed of 425 mph at 20000 feet. All internal fuel tanks were eliminated from the wing in the F4U-1D.

VOUGHT XF4U-3 (MFR.) The Navy early in World War II was concerned about lack of a very high altitude fighter, and in March 1942 contracted with Vought to produce a Corsair with a turbosupercharged Pratt and Whitney R-2800 Double Wasp version developing the full 2000 horsepower at 30000 feet and providing a speed of well over 400 mph at that altitude. This was the XF4U-3 version shown in the photo with belly air scoop, one of three F4U-1 conversions. The project was not a success and luckily the requirement for a very high altitude Navy fighter did not develop during the war.

VOUGHT F4U-4 (MFR.) This Vought photo of their spin demonstration airplane dated January 5, 1945 shows the F4U-4 Corsair, one of over 1400 ordered in January, 1944 and started into production the end of that year. Only a few got into Pacific combat areas late in the war in 1945. The -4 version had a new C model Double Wasp engine of 2100 horsepower dry in a new cowl design driving a four blade propeller, and a redesigned cockpit. The fighter could attain a top speed of over 440 mph at medium altitude and was considered by many to be the best flying of all Corsair models.

VOUGHT F4U-4N (MFR.) A small number of F4U-4N night fighter versions with APS-5 radar mounted on the starboard outer wing panel as shown in this company photo were built as an improvement over the early small number of F4U-2 night fighter conversions of the F4U-1. The aircraft pictured had six .50 caliber wing machine guns and rocket stand-offs under the wing.

VOUGHT F4U-4P (MFR.) A post World War II photo of one of the very few F4U-1P photo reconnaissance versions of the Corsair is shown at NATC Patuxent River, Md. attached to a land catapult. The only location available for cameras was in the belly of the aft fuselage, but no indication can be seen of such an installation in the photo.

VOUGHT F4U-4XB (MFR.) One of the two original 1943 prototypes for the F4U-4 Corsair, these converted F4U-1 aircraft designated F4U-4X as seen on the cowl of the aircraft in flight. The main change from the -1 was installation of the new Double Wasp version with the characteristic lower lip intake.

VOUGHT F4U-5 (Author) The next version of the Corsair was the F4U-5, a civil model of which is shown in a 1980 photo at NAS Willow Grove, Pa. Many -5s were purchased with the initial aircraft first flying after the war in the spring of 1946. A total of 328 F4U-5s were produced using Fiscal Year 1947 funds, 75 of them -5N night fighters, and 30 being -5P photo planes. In the next three fiscal years a total of 240 more F4U-5N aircraft were provided for with radar pods on the right wing. Production ended in 1951. The -5 had lower cheek cowl inlets for a new version of the R-2800 providing 2300 horsepower, a further cockpit rearrangement, armament of four 20mm cannon, a centerline pylon for a 2000 pound bomb along with the two inner wing pylons and eight rocket mounts. The F4U-5P had camera doors in the aft fuselage. High speed was about 460 mph.

VOUGHT F4U-7 (MFR.) Although not purchased for use by the US Navy, the F4U-7 was given a Navy designation, and as the last of the Corsair line is pictured in flight with the markings of the French Navy, to which 94 aircraft were delivered in 1952. The Corsair was the last US propeller fighter. The F4U-6 model was redesignated as an AU-1 attack plane.

VOUGHT XF5U-1 (MFR.) An unusual aircraft configuration based on the design of NACA engineer Charles Zimmerman, the XF5U-1 "Flying Pancake" was supposed to have a very wide speed range from landing at about 20 mph to a high value of about 450 mph. Powered with two Pratt and Whitney R-2000 piston engines each of 1600 horsepower both connected through shafting to two large propellers, the 8.5 ton gross weight fighter never got to fly. A flying model of the configuration, the V-173, made a first flight in late 1942. The XF5U- was completed at the end of World War II, but less than two years later the Navy stopped the project.

VOUGHT XF6U-1 (MFR.) The first Vought design for a jet fighter, the XF6U-1 Pirate pictured was one of three single seat prototypes ordered in December, 1944 using a 3000 pound thrust Westinghouse J34 engine. The aircraft featured Metalite sandwich panel construction for stiffness and smooth contours. First Pirate flight took place in October, 1946. The original flights were made without an afterburner; later one was added to increase jet thrust by about 1000 pounds for short periods. Armament was four 20mm cannon in the nose as shown by the faired bulges in the photo.

VOUGHT XF6U-1 (MFR.) The company photo shows an XF6U-1 Pirate with the original tail configuration before the addition of afterburner and more tail area flying over Muroc Dry Lake on March 5, 1947. The Navy ordered 30 production F6U-1 fighters in February, 1947, but severe prototype development problems were occurring, principally due to instabilities and marginal engine thrust. The production aircraft of 1949-50 still encountered problems to the extent the Pirates were never put into squadron service.

VOUGHT XF7U-1 (P.Bowers) A highly unusual single place twin jet Navy swept wing fighter with no horizontal tail (ailerons performed the elevator function and became ailevators), and designed to use afterburners from the start, the XF7U-1 Cutlass was initially ordered as three prototypes in mid-1946 and later in mid-1948 14 F7U-1 aircraft were also ordered. Though the aircraft shown has been called an F7U-1 by Navy number the label XF7U-1 is clearly marked on the fuselage. First flying in 1950 the F7U-1 development program was delayed by technical problems and fatal airplane crashes. A new version, the F7U-2 was cancelled and no F7U-1s reached operational service.

VOUGHT XF7U-1 (MFR.) The photo shows the XF7U-1 Cutlass configuration well. The nose boom is for flight test purposes. Powered by two Westinghouse J34 jet engines with afterburners having a maximum thrust of 4900 pounds each, the 10 to 12 ton gross weight fighter had a high speed of just under 700 mph at low altitude. Armament consisted of four 20mm cannon low in the fuselage nose. Dive brakes were inboard on the wing trailing edge. Flight testing of this unusual aircraft was a disaster. Two of the F7U-1s went to the Blue Angels.

VOUGHT F7U-3 (H.Gann) A December, 1955 photo, taken ten months after delivery of 180 fighters was completed, shows the much-revised F7U-3 Cutlass. Everything was redesigned except the general configuration. The photo shows the method of wing fold, underwing store station placement, a nose with APQ-30 radar, and 20mm cannon ports just above the engine air intakes. Also illustrated is the slat on an outer panel leading edge and the long nose wheel strut to give the required high angle of attack for takeoff.

VOUGHT F7U-3 (MFR.) An early test F7U-3 is shown in flight where differences from the F7U-1 can be noted. The -3 model was originally ordered in the summer of 1950. Chances are good this aircraft was powered by two Allison J35 engines because several early -3s came out before the intended Westinghouse J46 engine was available. The Cutlass had one of the longest development periods of any aircraft; it was over five and one half years between F7U- contract award and operational service for the F7U-3.

VOUGHT F7U-3 (MFR.) A Vought publicity photo of F7U-3 Cutlasses in close formation was intended to show the -3s in service. The Cutlass served initially in only four Navy squadrons, later increased to eleven, and was out of service by the late 1950s. Powered by two Westinghouse J46 jet engines with about 6000 pounds of afterburning thrust each, the 15 ton F7U-3 had a high speed of about 690 mph at low altitude.

VOUGHT F7U-3M (P.Bowers) The missile-carrying F7U-3M Cutlass, 98 of which were bought by the Navy by 1955, is shown in the photo intended for an attack squadron as shown by the VA- on the tail. One of the earliest missile fighters, this Cutlass retained the four 20mm cannon above the air intakes but could carry four Sparrows. Wing leading edge slats and long dual wheel nose gear are shown well. The pilot had very good visibility from his high forward location.

VOUGHT F7U-3P (MFR.) Flight photo of an F7U-3P photo reconnaissance version of the Cutlass shows the nose radar deleted and nose camera installations aboard. The gun armament was left out and more fuel was added. Only a dozen of this Cutlass variant were manufactured. It is doubtful that such a small number of aircraft were used operationally to any extent.

VOUGHT XF8U-1 (MFR.) After having real problems with their first two Navy jet fighters, the XF6U-1 and F7U-1 and -3, Vought really got it right with their F8U- Crusader series. The photo shows one of two XF8U-1 prototypes built against a mid-1953 Navy contract after Vought had been declared competition winner. A single seat single engine swept wing fighter powered with an afterburner-equipped Pratt and Whitney J75 jet engine, the most unusual feature of the XF8U-1 was a variable incidence wing.

VOUGHT F8U-1/F-8A (USN) Close view of an F8U-1 Crusader shows interesting details including the nose air intake. The variable incidence wing is in the low flush high speed position, details of the canopy are apparent, the ports for the 20mm cannon are near the head of the sergeant, there is a small gun ranging radar at the nose, the speed brake in the fuselage belly is partly down, and the arrangement of the main landing gear is shown. Also apparent is a stores mount alongside the fuselage. The flight refueling probe and the rocket launching tray are cleanly retracted. The 16000 pound maximum thrust J75 engine powered the 13.5 ton F8U-1 to a high speed of just over 1000 mph at altitude.

VOUGHT F8U-1/F-8A (H.Andrews) The first production version of the Crusader, an F8U-1 (later in 1962 redesignated F-8A) is shown from Navy Squadron VF-174 on board carrier Independence in 1960. The variable incidence wing is up and leading and trailing edge wing flaps are shown in the low speed condition. First flight of the F8U-1 took place in late 1955; it entered actual front line service in early 1957. Over 300 F8U-1 fighters were built.

VOUGHT F8U-1E/F-8B (H. Andrews) A Marine F8U-1E Crusader variant of 1958-9 is shown on the deck of the USS Independence in January, 1960 with the wing raised for takeoff. A total of 130 were built. This version, redesignated F-8B in 1962, had a limited all-weather capability by use of a new APS-167 radar in the nose. This F8U-1E running up is ready to move onto a catapult. Note the trailing arm type nose gear.

VOUGHT F8U-1P/RF-8A (MFR.) The photo reconnaissance model of the Crusader is shown in an early manufacturer's photo with a total of 144 built after first flying in late 1956. Armament was dispensed with and cameras were installed in the recontoured lower front fuselage. The photo shows camera ports on the side and underneath the fuselage looking forward and down. The wing fold joint is at the location of the leading edge sawtooth.

VOUGHT F8U-1P/RF-8A (USN) A Navy flight photo taken in April, 1961 of an F8U-1P, about a year after production ended. Unarmed, this Crusader model made an excellent high speed photo reconnaissance aircraft. The sawtooth break in the wing leading edge can be seen. The Crusader was a long airplane; wing span was only 35 feet eight inches while length was over 54 feet. The -1P airplane was used extensively in the Vietnam War.

VOUGHT F8U-1T/TF-8A (A. Schoeni) A two place trainer version of the Crusader is shown going through carrier qualification testing aboard the USS Independence in 1962. Note the FT for "flight test" on the nose and the port wing folded. Most noteable is the long canopy raised over the tandem cockpit arrangement. Only one -1T trainer was converted from a -1 airplane; no more trainer versions were built.

VOUGHT F8U-2/F-8C (UNK.) The length of the Crusader is emphasized in this photo of an F8U-2 of Navy Squadron VC-7. A total of 187 F8U-2 variants were delivered in the late 1950s and into 1960. The new feature easily distinguishing the -2 version from the -1 was the two long ventral fins aft on the fuselage as a correction for stability problems. Small scoops on the afterburner for cooling air were also new. The outline of the aerial refueling probe compartment can be seen forward of the national insignia.

VOUGHT F8U-2/F-8C (H.Andrews) Carrier photo of an F8U-2 Crusader shows the fighter with wings unfolding and in the "up" incidence position readying for takeoff. A Sidewinder missile is mounted alongside the fuselage. The Pratt and Whitney J75 engine generated about 17000 pounds of thrust in afterburner giving this 14 ton gross weight Crusader a maximum speed of about 1100 mph at high altitude.

VOUGHT F8U-2N/F-8D (USN) A Crusader ready to go! The photo shows an F8U-2N taking off a carrier deck. Appearing first in 1960 with 152 fighters turned out, this aircraft had an uprated J75 engine increasing speed by 100 mph at high altitude and more internal fuel replacing the rocket pack along with new avionics. Armament consisted of up to four Sidewinder missiles to supplement the four 20mm cannon.

VOUGHT F8U-2NE/F-8E (Author) An F-8E Crusader of Squadron VF-211, formerly an F8U-2NE until redesignated in 1962, at a 1973 airshow at NAS Willow Grove, Pa. The F-8E was the final Vought new production model, 280 of which were built in the early 1960s. The E version could be equipped with underwing store racks to hold a variety of bombs or missiles. Crusaders had considerable success in air battles during the Vietnam War.

VOUGHT RF-8G (USN) The RF-8G variant of the Crusader was a Vought-reconditioned RF-8A model with an uprated J75 engine, ventral fins added, and new avionics and camera installations. One camera side port of the RF-8A is blocked out on this RF-8G of VFP-306 at NAS Memphis in November of 1976. The RF-models were among the last Crusaders to be phased out of Navy service. They were still 1000 mph aircraft.

VOUGHT RF-8G (USN) Three RF-8G photo reconnaissance Crusaders show the broad swept wing planform of the aircraft with the one sawtooth on the leading edge which was also the position of the wing fold joint. There were over 70 RF-8Gs remanufactured from RF-8As with updates, starting a new business for aircraft companies which continues to this day. The practise is often much more economical than making totally new airplanes.

VOUGHT F-8H (J.Weathers) The H version of the Crusader is shown in the Naval Reserve Squadron VC-13 at NAS New Orleans. In the mid-1970s the Crusaders went from front line squadrons to the Reserves. The several dozen F-8Hs were re-manufactured by Vought from F-8Ds with engine and equipment updates. The wing fold is shown clearly. Nothing else on the airplane folded for stowage.

VOUGHT F-8H (MFR.) An unmarked F-8H fresh from the factory after re-manufacture is shown in a company public relations flight photo. A few H models apparently got to serve in Vietnam. Powered with a Pratt and Whitney J57 jet engine with 18000 pounds thrust in afterburner and a gross weight of almost 15 tons, the F-8H had a high speed still just over 1000 mph at altitude and over 700 mph down low.

VOUGHT F-8J (R.Besecker) Well over 100 F-8E Crusaders were redone by Vought to make almost-new F-8J fighters. Shown in a March, 1970 photo is an F-8J of Navy Squadron VF-191 from carrier Oriskany. Equipped to carry bombs, missiles, or drop tanks, the J aircraft got to 16-18 tons gross weight. When this photo was taken the Crusader was still a few years away from retirement.

VOUGHT F-8L (R.Besecker) A lineup of Marine Squadron VMF-321 F-8L Crusaders looks like the aircraft have been well used. A total of 63 F-8L fighters were updated F-8Cs. Note that these aircraft were not equipped with the aft ventral fins or the afterburner air scoops, nor is there any indication of wing or fuselage mounts for missiles.

VOUGHT F8U-3 (MFR.) An all weather fighter design resembling the earlier F8U- series in a general way, but actually a much different aircraft, the F8U-3 had a new airframe, a different engine, and new equipment and systems. The -3 was a competitor to the McDonnell F-4 Phantom. Both prototypes had a first flight in 1958 and a subsequent successful flight test program with a third airplane starting test flights, but by the end of 1958 the McDonnell aircraft had won the fighter competition and the Pratt and Whitney J75 powered F8U-3s were turned over to NASA for research. The photo shows the new engine inlet design and one of the missiles carried. No guns were aboard.

VOUGHT F8U-3 (MFR.) Another view of an F8U-3 prototype shows the forward-raked engine inlet design, the recessed stowage for missiles (the third missile was located under the fuselage), and the new aft ventral fins that could be lowered in flight to increase lateral-directional stability at very high speeds. Another feature was a new automatic flight control system. Powered by a J75 engine the 19 ton gross weight F8U-3 could fly well over 1400 mph at very high altitudes. Normal ceiling was over 50000 feet; the prototypes were actually zoom climbed to altitudes of over 60000 feet.

LOCKHEED XFV-1 (Mfr.) One of two single seat turboprop powered VTOL fighter design prototypes (the other being the Convair XFY-1), the Lockheed XFV-1 first flew in June of 1954 using a test conventional landing gear enabling a normal horizontal takeoff. In a series of 22 flights the XFV-1 was hovered only at altitude and never performed a true vertical takeoff or landing. Powered by an Allison T40 twin power section turboprop engine of about 5500 horsepower driving a 16 foot diameter dual rotation Curtiss Turbolectric propeller, the 16200 pound gross weight aircraft had an estimated top speed of over 550 mph. Gun or missile armament was to be located in the wingtip pods. The concept was later deemed not operationally feasible and was abandoned.

WRIGHT NW-1 (MFR.) Wright developed an unusual configuration for its Navy NW-1 (Navy Wright-1) racer designed for the 1922 Pulitzer Race where it was flown by Marine Lt. Sanderson. The engine failed and the aircraft went into a lake. A sesquiplane with the small lower wing well below the fuselage, the NW-1 was powered by a liquid cooled 750 horsepower Wright Tornado engine with Lamblin radiators mounted between the wings. The plane was completely redesigned to a more conventional biplane with skin cooling, this being an NW-2 intended for the 1923 Schneider Cup race as a twin float seaplane. An accident prevented the NW-2 from racing.

WRIGHT F2W-1 (MFR.) Like the NW-1 the Wright F2W-1 was actually a racer instead of a fighter, but was given a fighter designation (probably to cover for Navy funds used for racing). Two F2W-1s were built and tests were initiated in the summer of 1923 for the Pulitzer Race of that year. One aircraft ran out of fuel and crashed. The second plane was modified to a twin float seaplane for the 1924 Schneider Race but crashed after a first flight.

WRIGHT F3W-1 (MFR.) Contrary to the previous Wright racing planes, the Wright F3W-1 (or XF3W-1) Apache was a completely conventional biplane shown here as a landplane and was designed as a fighter. Intended to use either the Wright Whirlwind engine of 220 horsepower or a Wright P-1 Simoon engine of 325 horsepower, neither proved suitable and, probably to the anguish of Wright, a Pratt and Whitney supercharged Wasp engine of about 400 horsepower was used. In fact this aircraft was the first installation of the famed Wasp powerplant. The claim to fame of the F3W-1 was its altitude record of over 43000 feet piloted by Lt. A. Soucek.

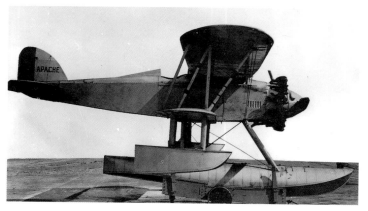

WRIGHT F3W-1 (MFR.) The F3W-1 was convertable to a single main and wingtip float seaplane as shown in this Wright photo. As a seaplane the F3W-1 established an altitude record of 38560 feet again flown by Lt. Soucek. The Wasp engine drove the 2120 pound aircraft to a top speed of about 160 mph as a landplane. Wingspan was 27 feet four inches. No more Wright Navy aircraft were built, and the airframe part of the Wright Aeronautical Corporation soon ceased to exist.

CONVAIR XFY-1 (P.Bowers) Like the Lockheed XFV-1 the Convair XFY-1 "Pogo" was another single seat turboprop tail-sitter VTOL fighter with a 16 foot diameter dual rotation Curtiss propeller. Because of the tail fin arrangement the Pogo had to make vertical takeoffs and landings, the latter being extremely difficult for the pilot. The XFY-1 is shown in an air show display and labeled the first vertical takeoff plane. After tethered tests inside the big blimp hangar at Moffat Field in California the XFY-1 made a first free flight from vertical takeoff through transition to conventional attitude and back to vertical landing in November of 1954.

CONVAIR XFY-1 (MFR.) The XFY-1 Pogo" sits on its four poster landing gear ready for takeoff with pilot Skeets Coleman on a tiltable seat. The long spinner covered the two hubs of the dual rotation propeller and a radar scanner up front ahead of the propeller, this latter supported by a nonrotating tube within the dual rotating shafts running back through the center of the reduction gearbox of the Allison T40 5500 horsepower twin power section turboprop engine. Delta wing tip pods were to house the armament.

CONVAIR XFY-1 (MFR.) The Pogo performs a vertical flight hover (*left*), literally hanging the 16250 pound gross weight aircraft on its propellers. The Allison XT40 had to be specially modified for vertical operation. The XFY-1 had an estimated top speed of about 600 mph and a wingspan of 27 feet eight inches. Pilot difficulty on landing vertically precluded use of the aircraft operationally and the Navy gave up on the tail-sitter VTOL concept.

CONVAIR XF2Y-1 (MFR.) In April, 1953 Convair test pilot E.D. Shannon made the first flight of an experimental delta wing twin jet seaplane fighter using hydroskis called the Sea Dart. The December, 1952 photo shows one of the prototypes (there were five aircraft, an XF2Y-1 and four YF2Y-1s) in the water; only three test aircraft were flown. Tests were made from 1952 to 1957 with two engine types and many twin and single hydroski designs; a single ski proved best for water operations. On November 4, 1954 the #2 aircraft, a YF2Y-1, crashed during a low altitude pass in a public demonstration. Final power installed was two Westinghouse J46 jets with 4000 pounds thrust normally and 6000 pounds each in afterburner giving the 22000 pound craft an estimated high speed of Mach 1.0 at 35000 feet. The program was canceled by the Navy in 1957 after about 450 test operations.

GENERAL DYNAMICS/GRUMMAN F-111B (R.Besecker) The photo shows the Navy version of the TFX (Tactical Fighter-Experimental) at Edwards AFB in May of 1968. Grumman had design responsibility for the Navy variant which was to share a high degree of commonality by dictate with the USAF F-111A version. This requirement was to be the undoing of the Navy aircraft and only seven were built.

GENERAL DYNAMICS/GRUMMAN F-111B (H.Andrews) Like its Air Force F-111A counterpart the F-111B was a two seat side-by-side swing wing twin turbofan engined fighter. This flight photo shows the wing in an intermediate sweep angle position. The first F-111B flew on May 18, 1965. It was demonstrated that performance as a carrier fighter was deficient and the program was a political football. Armament was projected to be six Phoenix missiles.

GENERAL DYNAMICS/GRUMMAN F-111B (MFR.) The F-111B is shown with the swing wing fully aft to a sweep angle of 72.5 degrees for high speed flight. In this regime roll control was effected by the aft stabilators working in a differential mode; at low speeds the wing swung forward to a 16 degree sweep and roll control was gained by spoiler action. The F-111B used two Pratt and Whitney TF30 turbofans each with 20000 pounds of afterburning thrust giving a high speed of about Mach 2.0 at altitude. The 70000 pound aircraft was overweight and not deemed carrier compatible, and the Navy program was cancelled in 1968.

NAVY LIGHT TRANSPORT AIRPLANES

During a short period of time in the late 1930s and through most of World War II the Navy used a VG class designator for small personal transport airplanes. These were not special Navy designs but rather off the shelf civil light single engine transports and sport aircraft procured or drafted into the service. Only three designs are of note, the Beech GB-, the Howard GH-1, and the Fairchild GK-1, these being respectively navalized versions of the Beech Model 17 Staggerwing, the Howard DGA-15 Damn Good Airplane, and the Fairchild Model 24. Used as light utility transports for personnel, a large number of Beech aircraft were employed; only a few of the Fairchilds or Howards got into the Navy.

BEECH GB-1 (H.Thorell) A Navy version of the popular civil Model D-17S utilized for various light personnel transport duties, ten of which were purchased with seven delivered in 1939 and the remainder in 1940. The one shown was used by the US Naval Attache in Mexico City. A few civil aircraft were later drafted in as GB-1s; two of these served in the Marines. Powered by a 450 horsepower Pratt and Whitney Wasp Junior engine, the 4200 pound gross weight "Staggerwing" aircraft had a high speed of about 210 mph and could cruise at 200 mph. at about 10000 feet.

BEECH GB-2 (MFR.) A company photo shows a production lineup of GB-2 Travelers. The GB-2 was a slightly modified follow-on to the GB-1 with 342 of these Navy and Marine Staggerwings put out during World War II from late 1941 into 1944 with a few of these going to the United Kingdom. Powered with an uprated militarized R-985 Wasp Junior engine of 450 horsepower for takeoff the GB-2 weighed 4250 pounds loaded and had about the same performance as the GB-1.

NAVY HOSPITAL AIRPLANES

Many Navy aircraft have been used as medical evacuation types, these usually being cargo or personnel planes in the VR or VC category having interiors fitted out with litter accomodations. The only aircraft found specifically designated as a hospital plane in the VH category that is known was the Loening XHL-1 of 1929 made as a revision of a basic observation amphibian.

LOENING XHL-1 (USN) Two 1929 hospital plane variants of the basic Loening OL- observation amphibians, the XHL-1s were seven place double bay biplane aircraft with a pilot sitting in an open cockpit forward of and above a closed cabin for up to six occupants. One Marine aircraft was lost in a 1929 crash. Powered by a 525 horsepower Pratt and Whitney Hornet engine, the 5600 pound gross weight XHL-1 had a top speed of about 120 mph and a wingspan of about 47 feet. The three bladed propeller just cleared the single "shoe-horn" main float.

NAVY UTILITY AIRPLANES

The Navy has employed a wide variety of airplanes in the VJ or utility class over the years. In some cases only one plane of a particular type was procured, and in other cases quantities similar to those of major combat aircraft have been used. Many times in the 1920s obsolete combat aircraft were used for general utility missions without a designation change, but these were not always suitable for many roles, and more specialized planes had to be procured.

Three aircraft types of the 1927-28 period designated by the Navy were the Ford Trimotor (XJR-1) as the Navy's first transport, the Fairchild XJQ-1 navalized version of the Fairchild Civil Model 71 convertable from wheels to twin floats as required, and the XJA-1 naval version of the Fokker civil Super Universal transport. All of these were simple conversions of popular civil types; the Navy had no money for specially designed aircraft in this role.

Two Waco biplanes were procured by the Navy in 1934 as the XJW-1 and initially used for trainers in the dirigible hook-on project.

By far the most numerous of Navy utility planes came from Grumman starting in 1934 and going well into World War II as the JF- and J2F- series of single engine biplane amphibians. The J2F- Duck series was similar to the JF-s except revised with a larger hull, and it was started in service in 1937. Used by Navy, Marines, and Coast Guard, these Grummans, later built by Columbia, were used in a variety of roles from photography to rescue, to ASW work.

In 1936 the Navy purchased two civil airplanes from Fairchild, first a single JK-1 navalized Model 45 cantilever low wing monoplane used as a command airplane, and then a few civil Model 24s as J2K-s for the US Coast Guard.

The next year, 1937, saw several civil aircraft designs acquired by the Navy as utility types. One was the first Beech Staggerwing as the JB-1, another the single Bellanca JE-1 light transport, the twin engined Lockheed Model 12 Electra Junior as the JO- series, one for testing the tricycle aircraft landing gear for possible carrier use, and the Stearman Hammond JH-1 twin tailed light pusher plane for radio control tests.

Grumman sold a large number of their civil G-44 Widgeon twin engined amphibians to the Navy in 1941 and through the war years as the J4F- series used in a myriad of utility duties.

A number of early model Martin Marauder bombers were obtained from the USAF and converted during wartime as JM-1 and JM-2 utility planes powerful enough to be used as fast target tugs.

A Grumman design for a follow-on to the JF- and J2F- Duck series was fabricated in prototype form by Columbia Aircraft in 1946 as a monoplane XJL-1 but there was no production after the war.

The most recent aircraft categorized by the Navy as in the VU utility class was the JD-1, the naval conversion of the USAF B-26 Invader light bomber. These few ex-USAF aircraft were used in various roles including drone director planes.

FOKKER XJA-1 (USN) A single civil Fokker Atlantic Super Universal was tested by the Navy in 1928 as a transport, in fact the A in the designation actually signified transport in an early system. An eight place high wing cabin monoplane, the XJA-1 was powered by a 420 horsepower Pratt and Whitney Wasp engine. The aircraft had a fabric covered welded steel tube fuselage, and the wing was an all wood structure. Wingspan was 50 feet eight inches and high speed of the 5150 pound gross weight cantilever wing plane was about 135 mph.

BEECH JB-1 (H.Thorell) The initial four to five place Navy Beech Staggerwing, a single 1937 JB-1 light utility transport for high level officers, was a C-17R (Serial #115) civil model equivalent used until 1939. The JB-1 was powered by a Wright R-975 Whirlwind engine of 450 horsepower for takeoff driving a Hamilton controllable pitch propeller. Gross weight was 3900 pounds and high speed was about 210 mph at sea level. The unique reverse stagger wing and I-strut biplane had a span of 32 feet for both wings and wing chord was five feet.

DOUGLAS JD-1 (P.Bowers) A 1964 photo of a Navy version of the Air Force B-26 Invader shows the typical clean lines of the Douglas aircraft, designated a JD-1 utility plane. A total of 140 USAF A-26Cs (later B-26Cs) were used by the Navy for various duties including target tugs. The JD-1 was powered by two Pratt and Whitney Double Wasp engines each of 2100 horsepower giving the 18 ton gross weight aircraft a high speed of almost 350 mph. Wingspan was 70 feet.

DOUGLAS JD-1D (P.Bowers) A modification of the Navy Invader was made in a few cases to a JD-1D version used to air-launch pilotless drone target vehicles such as the jet powered Ryan Firebee and later versions. The photo shows a drone controller Invader of Navy Squadron VU-3, one of at least four squadrons using the type for many years. The store under the port wing is apparently an external fuel tank.

BELLANCA JE-1 (MFR.) Like the Army in the 1930s the Navy was apt to order single examples of a popular civil aircraft type for utility service. The JE-1 was actually a single Model 31-55 six to nine place Bellanca Senior Skyrocket high wing monoplane delivered in 1937 and used a 550 horsepower Pratt and Whitney Wasp engine providing a top speed of about 185 mph at 5000 feet. Gross weight was 5600 pounds and wingspan 56 feet. The aircraft was stationed for a time at NAS Anacostia.

GRUMMAN JF-1 (H.Thorell) The initial version of the famous Grumman Duck three place biplane amphibian (so named later); one of several assigned to Navy Squadron VS-3 for utility duties aboard carrier Lexington in 1934 equipped as required with a tail hook. First flight took place on April 17, 1934 and delivery of 27 aircraft continued from May, 1934 to January, 1935. Using a 700 horsepower Pratt and Whitney Twin Wasp engine driving a three blade Hamilton propeller the 5400 pound gross weight JF-1 had a top speed of 168 mph. Armament consisted of one rear .30 caliber gun and two 100 pound bombs. Span was 39 feet.

GRUMMAN JF-2 (USCG) A version of the Duck for the US Coast Guard first flown in October, 1934 with 15 aircraft produced from that month until October, 1935. One plane went to the Marines. Like the JF-1 in many respects, the JF-2 used a 700 horsepower single row Wright Cyclone, had no armament, and incorporated RDF equipment. A JF-2 claim to fame was establishment of a single engine amphibian speed record of 191 mph in December of 1934. Gross weight was 5500 pounds. The Argentine government bought eight additional aircraft.

GRUMMAN JF-3 (H.Thorell) The photo shows the JF-3 as a Navy version of the JF-2 with a 750 horsepower Cyclone engine driving the same Fay Egan propeller (three bladed to get enough blade area to absorb the power and still clear the float diameter-wise). Only five JF-3s were produced, these in September and October of 1935. The aircraft went to Navy and Marine reserves at several bases without tail hooks. Armament was like that of the JF-1. High speed was 175 mph; gross weight 5600 pounds.

GRUMMAN J2F-1 (H.Thorell) Essentially an improved JF- type aircraft, the J2F-1 was produced to the extent of 29 aircraft from early April to mid-June of 1936. Equipped with a carrier hook and catapulting gear along with more equipment bringing gross weight up to about 6100 pounds, the J2F-1 had a 750 horsepower Wright Cyclone engine driving a nine foot diameter propeller and a top speed of just over 170 mph. One aircraft was revised with full span upper wing flaps in early 1937 and called a J2F-1A.

GRUMMAN J2F-2 (USN) A Navy photo shows the J2F-2 version of the Grumman Duck on October 7, 1938, four months after delivery of the first of 30 aircraft produced. J2F-2 production ended in mid-November of that year. Again powered with the Wright R-1820 Cyclone engine of 750 horsepower the 6180 pound gross weight Duck had a high speed of 168 mph and could be armed with one fixed and one flexible .30 caliber machine gun along with two 100 pound bombs.

GRUMMAN J2F-2 (USN) Flight photo of the J2F-2 shows the main wheel retracted into the float. Although the Duck was normally called a three place aircraft there was room for an extra man in the capacious fuselage. Most of the 30 J2F-2s went to the Navy, but nine were operated by the USMC.

GRUMMAN J2F-2A (H.Thorell) An example of a Marine J2F-2A amphibian sits painted with the 1940 neutrality star on the forward fuselage. The A in the designation signified an armed aircraft with a fixed and a flexible gun. Bomb racks were under the lower wing. Most of the J2F-2s became J2F-2As for the USMC when armed in 1939. Some served in the US Virgin Islands. Wing span was 39 feet; length was 34 feet.

GRUMMAN J2F-4 (H.Thorell) After 20 J2F-3s had been produced from February to June of 1939 for VIP usage Grumman switched to a J2F-4 model Duck first flown by Grumman test pilot Converse on September 11, 1939. Delivered from mid-September to June, 1940, 32 aircraft of the -4 variety were produced. The photo shows a Duck assigned to the NAS at Seattle, Washington. The window in the lower mid-fuselage can be clearly seen.

GRUMMAN J2F-4 (USN) Shown only 12 days after delivery of the first airplane is a J2F-4 ready for assignment to a Naval Air Station, as were all of this batch of Ducks. A very useful aircraft, the J2F-4 Duck was equipped for such tasks as smoke-laying and target towing. It was powered by a 790 horsepower Wright Cyclone driving a three blade nine foot diameter Hamilton propeller giving the 6330 pound gross weight amphibian a high speed of about 175 mph.

GRUMMAN J2F-5 (P.Bowers) An export version of the J2F-5 for Mexico is shown. Major Grumman production of the Duck started just before World War II with 144 of the utility amphibians turned out from late August, 1941 to March, 1942 after a first flight on July 10, 1941. Using an uprated 950 horsepower Cyclone engine in a longer chord cowling and a Hamilton constant speed propeller, the J2F-5 (one of which went to the Army as the OA-12) had a flexible .30 caliber gun aft and could lay smoke, tow targets, and could carry two light bombs or two 325 pound depth charges.

GRUMMAN J2F-5 (USN) Like many other aircraft types the Grumman Duck was used on anti-submarine patrol early in World War II. Shown in a Navy photo of April 23, 1942 a J2F-5 of Navy Squadron VJ-2 with the striped tail of that time carries bombs underwing and a .30 caliber gun at the ready. The J2F-5 model was to be the last produced by Grumman with this task taken over by Columbia Aircraft.

COLUMBIA J2F-6 (R.Besecker) A photo taken at the Reno Air Races in Nevada in the 1970s shows a restored J2F-6 built by Columbia Aircraft of Valley Stream, Long Island, N.Y. starting in 1943 after production was handed over by Grumman. Note that the J2F-6 designation and the airplane number are improperly placed, the designation belonging on the rudder.

COLUMBIA J2F-6 (H.Martin) A Grumman publicity photo shows a fine lineup postwar of Grumman-designed amphibians with a Columbia J2F-6 followed by a Goose, Widgeon, and a Mallard. Columbia put out 13 Ducks in 1943, 198 in 1944, and 119 in 1945. They were used as target tugs, ASW planes, photo planes, and for myriad other utility duties.

COLUMBIA J2F-6A (R.Besecker) A Duck assigned to NAS Atlantic City, N.J. is shown. The A stood for a plane capable of being armed. Power was supplied by a 1050 horsepower Cyclone engine giving the 7765 pound gross weight J2F-6A a high speed of 190 mph at 14000 feet with bombs aboard. The Grumman/Columbia Duck was a slow aircraft that seemed to take forever to climb, but proved a useful utility type.

GRUMMAN J4F-2 (MFR.) A lineup of three to five place J4F-2 Widgeon utility amphibians are shown ready for delivery at the factory. After 25 J4F-1s were produced for the US Coast Guard between July, 1941 and June, 1942 with a litter hatch and a bomb or 325 pound depth charge under the starboard wing Grumman launched immediately into production of a Navy J4F-2 version without the litter hatch. A total of 131 unarmed J4F-2s were delivered through World War II starting in July, 1942. Two went to the Marines and several to Britain. One USCG J4F-1 sank a German submarine off the Louisiana coast in August of 1942.

GRUMMAN J4F-2 (Edo) An Edo Corporation photo shows a Grumman Widgeon modified for special hull tests. Standard J4F-2 Widgeons used two Ranger L-440 engines each of 200 horsepower driving two blade wooden Sensenich propellers providing the 4525 pound gross weight aircraft with a maximum speed of about 150 mph. The J4F- amphibians were minor modifications of the civil Grumman G-44.

STEARMAN-HAMMOND JH-1 (H.Thorell) This off-the-shelf Stearman-Hammond Y-1S contestant for an every-man's airplane in 1935 was bought by the Navy to the extent of two aircraft in 1937. The JH-1 version first flew in December of that year and was used for experiments on radio control due to its safe flight characteristics. The twin boom pusher aircraft was powered by a 150 horsepower Menasco engine. Gross weight was 2150 pounds and high speed 130 mph.

FAIRCHILD JK-1 (H.Thorell) In 1936 the Navy purchased one civil Fairchild 45A five place light transport as a JK-1 model for carriage of high level naval officers; two more civil 45s were later drafted as JK-1s. Procurement of sample civil aircraft was common with both services pre-war. Powered by a 320 horse-power Wright Whirlwind engine the two ton gross weight plane had a high speed of about 165 mph.

FAIRCHILD J2K-1 (H.Thorell) Another Fairchild civil design, the three to four place 24R, was purchased for the US Coast Guard in 1936. Four aircraft, two J2K-1s and two J2K-2s with minor differences were assigned to Coast Guard bases as staff utility planes. Engine power was the air cooled Ranger of 145 horsepower. Gross weight was about 2500 pounds and high speed about 130 mph. As shown in the photo, these Fairchilds were probably the only USCG planes to use wheel pants.

COLUMBIA XJL-1 (H.Andrews) Designed by Grumman as a G-42 tricycle gear monoplane update of the Duck amphibian as early as 1939, pressure of other contracts forced the company to delegate work on the XJL-1 to Columbia Aircraft of Valley Stream, Long Island. First flight finally took place post-war on October 26, 1946; the second of two aircraft ordered flew several weeks later. Tested at NAS Patuxent River and then stored at Norfolk, the two planes were surplused in 1959 and sold privately. The amphibian had a six place passenger or cargo compartment inside the fuselage. A Wright R-1820 Cyclone engine drove the 13000 pound monoplane amphibian to a high speed of about 175 mph.

MARTIN JM-1 (USN) The Navy's version of a USAAF medium bomber, the Martin B-26 Marauder, is shown as a JM-1 utility aircraft in a flight photo of April, 1944. The USAAF serial number, 41-35600, is still on the vertical tail identifying the plane as a former AT-23B bomber crew trainer. Over 200 were used by the Navy as utility aircraft including target towing duties. Powered by two 2000 horsepower Pratt and Whitney Double Wasp engines, the 19 ton gross weight aircraft had a high speed of about 285 mph.

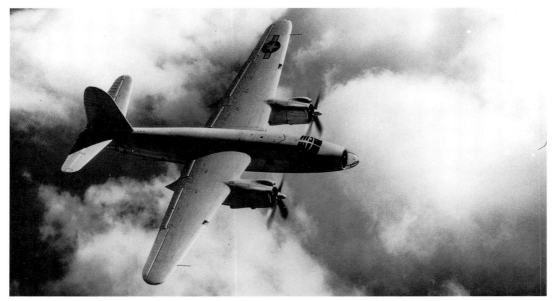

MARTIN JM-2 (H.Andrews) Stripped of armament and armor, and used for utility duties such as target tugs, one of about 50 JM-2 Navy Marauders is shown in an attractive flight photo. The -2 differed from the -1 in being a conversion of a later model Army B-26 bomber crew trainer. Performance and weights were very similar to the earlier model.

LOCKHEED JO-1 (H.Thorell) A single Lockheed 12A twin engined and twin tailed seven place Electra Junior was procured by the Navy with delivery in mid-1937. A sleek modern aircraft for its time, the plane was designated a JO-1 and used as a high level personnel transport. Power was supplied by two Pratt and Whitney Wasp Junior engines of 400 horsepower each providing a high speed of about 215 mph for the 8400 pound gross weight JO-1. The Electra Junior wingspan was 49 feet six inches.

LOCKHEED JO-2 (H.Thorell) Another Navy equivalent of the 12A Electra Junior was the JO-2, five of which served starting in 1937. The JO-2 shown and one other served the Navy; the others were used by the Marines as headquarters staff transports with one aircraft lost in the December 7, 1941 Japanese Pearl Harbor attack. Specifications were essentially the same as the JO-1 but cabin arrangements differed.

LOCKHEED XJO-3 (Mfr.) Another Electra Junior went to the Navy as an XJO-3 distinguished by employing a fixed tricycle landing gear as shown in the photo with the tail wheel retained as a tail down landing bumper. With the advantages of tricycle gear apparent the Navy wished to test carrier compatibility of such an arrangement which of course was eventually used on so many Navy aircraft. Power was the same as the other JO- types; gross weight was up about 250 pounds net, and because of the fixed gear high speed went down to about 175 mph.

FAIRCHILD XJQ-1 (USN) An NAS Anacostia, Md. photo of April, 1928 shows another popular civil aircraft procured as a single item for utility use, the Fairchild FC-2 conventional five place high wing monoplane of the period with a service designation of XJQ-1. Re-engining it provided an XJQ-2 designation. Later another Fairchild monoplane, a civil Model 71, was procured as an XJ2Q-1. Data for the XJQ-1 includes a Wright Whirlwind 7 of 220 horsepower, a wingspan of 44 feet, a gross weight of just over 3600 pounds, and a maximum speed of about 110 mph.

FORD XJR-1 (H.Andrews) With the big letters on the fuselage and the airplane number large on the tail there was no mistaking this Ford 4-AT Trimotor as a Navy airplane! This was the XJR-1 with the photo taken on April 20, 1928 at NAS North Island, San Diego, Ca. shortly after delivery to the Navy. Powered by three 300 horsepower Wright Whirlwind engines and with a gross weight well over five tons, the single big 68 foot span Trimotor rattled along at a high speed more than 110 mph for over two years in Navy service.

FORD JR-3 (H. Andrews) After two more Whirlwind powered JR-2s served with the Marines starting in 1929 Ford provided the Navy with three Pratt and Whitney Wasp powered Trimotors of the larger civil 5-AT design the next year, two of the aircraft going to the USMC. The photo shows a Navy aircraft as is made obvious by the large wing lettering. Pants are on the landing gear, but the 450 horsepower Wasps are uncowled. These all metal transports weighed about 12700 pounds loaded and had a high speed of about 135 mph with 450 more installed horsepower than the earlier Fords.

WACO XJW-1 (P. Bowers) The navalized Waco UBF three place sport biplane shown in the photo was one of two XJW-1 models procured in 1934 as trainers and utility planes for the Naval Heavier Than Air unit of airship USS Macon. As such it was equipped with an airship hook in its earlier days. With the Macon lost the XJW-1, minus hook, was assigned to NAS Anacostia as shown in the photo as a utility airplane where it stayed until World War II. Using a Continental R-670 engine of 210 horsepower, the 2300 pound gross weight plane had a top speed of about 130 mph.

NAVY UTILITY TRANSPORT AIRPLANES

The US Navy employed a utility transport category of aircraft from the late 1930s until a short time after the end of World War II. Of the six airplane types carrying the utility transport designator all were multi-engined with some transport capability and one was four engined. Two were land-based airplanes, three were amphibians, and one was a very large flying boat.

The first aircraft in the VJR series was a naval version of the twin engined Sikorsky S-43 amphibian labeled a JRS-1. Used by both Navy and Marines in small quantities, the clean lined JRS-1s entered service in 1937.

In 1939 a utility transport type used in large numbers entered service, first as an XJ3F-1 then redesignated as the JRF- Goose, the naval version of the Grumman G-21 civil amphibian. The aircraft had a long life and served for some time after World War II.

Another utility transport produced in quantity was the ubiquitous civil Beech Model 18 in naval garb as the various JRB- models. These were closely related to the Beech SNB-s in the scout training category but used primarily as light utility transports instead of trainers. The JRB-s started service in 1940 and lasted through and after the war.

The Cessna T-50 light civil transport was also used by the Navy as a JRC-1 twin engined Bobcat transport, though less so than the twin Beech. The Army Cessna counterpart was the UC-78.

By far the largest aircraft in the VJR class was the huge Martin JRM- series of flying boats originally derived from the XPB2M-1 patrol bomber prototype which served for many years starting midwar and continuing after for a time with Transport Squadron VR-2. Four of the JRM-1s had Wright R-3350 engines; one JRM-2 had Pratt and Whitney R-4360 Wasp Majors. Many records were set by the JRM- boats.

The last aircraft design in the VJR category was the Grumman XJR2F-1, two prototypes of which appeared in 1948. From these prototypes came a long line of Navy UF- and Air Force SA-16 Albatross amphibians used for a great variety of missions.

No more types were put in the VJR category; later aircraft would simply be characterized as transport planes in the VR category, and they would generally be landplanes.

BEECH JRB-1 (Mfr.) A 1940 five place naval version of the Beech B18S light twin engined aircraft, the JRB-1 utility transport came into the Naval inventory with orders totaling 11 aircraft. The JRB-1 was distinguished by a bulged upper cockpit area. They were similar to the Army F-2 photo plane version in having a hatch within the cabin door to allow an oblique camera installation. The aircraft were employed as utility types including photo planes and as radio controlled drone director planes. Engines were Pratt and Whitney Wasp Juniors of 450 horsepower powering the 7850 pound gross weight plane to a high speed of 210 mph.

BEECH JRB-2 (Mfr.) Another version of the Beech civil 18-S and generally equivalent to the Army's C-45A, the JRB-2 cockpit area was conventional with no upper bulge. The Navy ordered 15 of this model in 1940-1941 with most being used as light five passenger command transports for senior naval officers. A later 23 JRB-3s were built as the Navy equivalent of the Army C-45B model and some were used as photo planes. Performance and weight characteristics were quite similar to those of the JRB-1. Engines were still Wasp Juniors of 450 horsepower each.

BEECH JRB-4 (USN) The photo shows a 1943-44 seven place Beech JRB-4 Expeditor based at NAS San Diego, Ca., the first Army-Navy standardized military model of the light transport with the Army equivalent the UC-45F. A total of 328 Naval JRB-4 versions were produced. Again powered by two Pratt and Whitney R-985-AN-1 Army-Navy standardized Wasp Junior engines of 450 horsepower each, the 8725 pound gross weight JRB-4 had a high speed of 206 mph.

BEECH JRB-6 (P.Bowers) A post-war version of the Beech Expeditor based at NAS Norfolk, Va. is shown in a runup photo, this aircraft one of many JRB- types remaining after World War II and put through a 1951 factory modification program to modernize all elements of airframe, powerplant, and systems. Overall specifications of power, weights, and performance were similar to earlier versions.

CESSNA JRC-1 (Author) A postwar (1981) photo at an airshow of a rebuilt UC-78 airplane shows it painted up as a Navy JRC-1 Bobcat, actually civil registration N266C. Wartime UC-78 production went over 3000; 67 of these were provided to the Navy as JRC-1 light four place personnel transports. Powered by two 450 horsepower Jacobs engines, the 5700 pound gross weight Bobcat had a top speed of about 175 mph.

GRUMMAN JRF-1A (USN) After a single prototype XJ3F-1 Goose amphibian (named later) prototype was delivered in September, 1938 the Navy received five six seat JRF-1A aircraft as shown in this photo of September 19, 1939, five days after the first flight. The fifth JRF-1A was delivered in January of 1940. These amphibians were intended as target tow and photo planes with a camera hatch in the hull bottom. Starting in late November, 1939 an additional five JRF-1 Goose versions were delivered, these without the camera hatch. Powered by two Pratt and Whitney Wasp Junior engines of 450 horsepower each, these four ton gross weight amphibians could reach a high speed of 200 mph.

GRUMMAN JRF-2 (H.Thorell) This early version of the Goose was equipped for US Coast Guard operations such as air-sea rescue as was the later JRF-3. First flown on July 7, 1939, seven -2 and three -3 versions were built and delivered through late 1940. Six seats were provided or two litters could be fitted, and the JRF-3 was equipped with an autopilot. Twin Wasp Junior engines gave the JRF-2 and JRF-3 amphibians just about the same performance as the JRF-1s.

GRUMMAN JRF-5 (Walsh) One of the major production versions of the Goose amphibian is shown in December, 1949 at Alameda, Ca. A total of 184 of the JRF-5 aircraft were delivered from Grumman from July, 1941 throughout the war until September, 1945. Installed power, weights, and performance figures were similar to those of earlier versions. The six-seater could also carry two 325 pound depth charges , and tow target equipment could be installed. Some JRF-5s went to the Army as OA-13B observation amphibians, others went to the British.

GRUMMAN JRF-5G (UNK.) The photo shows an Alaska-based JRF-5G Goose, one of two dozen JRF-5s supplied to the US Coast Guard and differing only in specialized equipment. These aircraft were equipped with Army-Navy production-standardized Pratt and Whitney R-985-AN-12 Wasp Junior engines each of 450 horsepower. Wingspan was 49 feet.

GRUMMAN JRF-6B (USN) A Grumman JRF-6B with a British roundel on the fuselage sits for a Navy photo on February 6, 1942. These Goose amphibians were built under Lend-Lease for the British, and 50 of this version were produced from January, 1942 until March of 1943. Called the Goose IA by England, the planes were used as utility transports and navigational trainers. They could also be equipped with two 325 pound depth charges underwing for ASW work. Five of these stayed with the US Army as OA-9GR observation amphibians. Specifications were similar to other JRF- aircraft.

GRUMMAN XJR2F-1 (Mfr.) During late World War II Grumman started design of a larger replacement for the JRF- Goose series, and a contract was gained for two prototype aircraft, one of which is shown in the photo as the XJR2F-1 utility transport amphibian. These aircraft were to start a long line of successful service types for the USAF and USCG as well as the Navy. First XJR2F-1 flight took place on October 7, 1947 with both prototypes delivered to the USN by May 11, 1948.

GRUMMAN XJR2F-1 (Mfr.) Flight photo of the Navy XJR2F-1 amphibian shows clean lines. A patrol version with armament designated PF-1 was proposed, but not proceded with. Powered by two Wright R-1820 Cyclone engines of 1425 horsepower each driving three bladed Hamilton Standard propellers, the aircraft had a high speed of about 225 mph. Wingspan was 80 feet and length 62 feet. These prototypes were the forerunners of the later Navy UF-1 and Air Force SA-16A Albatross amphibians.

MARTIN JRM-1 (Mfr.) The Martin JRM-1 was a developed single tail version of the XPB2M-1R transport flying boat shown in the photo, itself a modification of the XPB2M-1 Mars long range patrol bomber first ordered as a prototype in August, 1938 and flown first in July of 1942. In 1943 the Mars was converted to a transport flying boat. Its wartime success in carrying heavy loads over long distances resulted in a 1944 order for 20 JRM-1 single tail cargo versions. Powered by four Wright R-3350 engines of 2200 horsepower each the approximately 72 ton gross weight flying boat had a top speed of over 220 mph.

MARTIN JRM-2 (USN) The order for 20 JRM- aircraft was reduced to six after World War II ended, one of these being the JRM-2 model shown in a Navy flight photo of May 10, 1948. The JRM-2 was powered by four 2300 horsepower Duplex Cyclones and had a maximum gross weight of 165000 pounds. The JRM- models all had single tails.

MARTIN JRM-3 (P.Bowers) The immense size of the Mars type flying boat is shown in the photo of a JRM-3 up on land using its beaching gear. The five JRM-1 aircraft were updated to JRM-3s. Mars boats were operated for several post-war years establishing a few flight records by west coast Navy Transport Squadron VR-2. The greater flight efficiency of large cargo landplanes was being established, however, and the Mars boats soon disappeared from the Navy. The JRM-3 was powered like the -2; high speed was about 225 mph.

SIKORSKY JRS-1 (H.Thorell) The JRS-1 amphibian was a Navy version of the civil Sikorsky S-43 design of which several were sold to airlines. The US Navy ordered 11 JRS-1 versions in 1937-8 of which two went to the Marine Corps. Eight were operated by the Navy's Utility Squadron VJ-1 and the operating Marine Squadron was VMJ-1. Normal seating was for 16 passengers with a crew of three. The JRS-1 was an exceptionally efficient amphibian type with good flying and water handling characteristics.

SIKORSKY JRS-1 (P.Bowers) One of the two USMC JRS-1 aircraft operated by VMJ-1 is shown in the photo. Main and tail wheels were hydraulically retractable. Two Pratt and Whitney R-1690 Hornet engines each of 750 horsepower were installed on the high wing of this 19100 pound gross weight amphibian giving the Sikorsky a high speed of 190 mph at 7000 feet. Wingspan was 86 feet and length 51 feet two inches.

NAVY TRAINER AIRPLANES

This section illustrates Navy trainers used or tested in the period approximately from the end of World War I through World War II, covering generally the years 1919 through 1947. Later model trainers are covered under the T category.

A large number of seaplane models were used for training of Navy fliers prior to and during World War I starting with Curtiss hydros, Curtiss F flying boats, Curtiss Model R seaplanes, and early Curtiss N-9 models with the Curtiss engine. Starting after the war in 1919 there were small numbers of N-9Hs and Vought VE-7s employed for training, mainly at Pensacola, Fl. The N-9H was the principal model used into the mid-1920s.

In 1924 the Naval Aircraft Factory put out some gunnery trainers as TG- airplanes. Also in the early 1920s an evaluation was made of several training seaplanes as an N-9H and VE-7 replacement, including NAF, Huff Daland HN-, Martin N2M-, and Boeing designs. The Boeing NB-1 of 1925 was selected for production, but the aircraft had a tendency towards flat spins causing several accidents. It was used with both liquid cooled and air cooled radial engines.

In 1926-28 there were several trainer designs tested by the Navy including an XN2B-1 Boeing follow on to the NB- series doomed to lose because of a powerplant problem, the big Curtiss N2C- Fledgling used in the Curtiss Flying Service which gained some production orders, and the Keystone XNK-1, which also gained a small Navy order.

In 1926 the first design of Consolidated, an NY-1 trainer, like earlier models convertible from landplane to seaplane, and somewhat similar to the Army's PT-3 primary trainer, was purchased. These Wright Whirlwind engine powered trainers in several variants were acquired over the next several years into the 1930s. The XN2Y-1 and XN3Y-1 from Consolidated in 1929 were not as successful.

In 1930 a very few New Standard NT-1 trainers with a Kinner engine were purchased based on the Civil Model D-25, but there was no large production. Similarly in 1931 a very few Consolidated N4Y-1s were procured for the US Coast Guard.

In 1935-36 two lines of Navy primary trainer aircraft were started which were to be the last of the mass-produced two seat biplane types. The first was the Stearman NS-1 of 1935, invariably used as a landplane, and the Naval Aircraft Factory N3N- of 1936 which was convertable to a seaplane. The Stearman line was used by both services, and the N2S- follow-ons to the NS-1 were put into mass production and utilized throughout World War II and beyond. The NAF N3N-3 models were the last biplanes in Navy service and were used as seaplanes at Annapolis.

The first low wing Navy trainers came out in 1937 as North American NJ-1s. These were cantilever monoplanes with fixed panted landing gear and enclosed tandem cockpits, and were similar to the Army's BT-9B basic trainer. Some were used as trainers and some as command transports.

A variety of Navy trainers came out in the early US pre-war years of 1940 and 1941 including many Howard DGA-15 aircraft used as NH-1 instrument trainers, an XN5N-1 prototype from the NAF, a large number of 1940 Spartan civil trainers given an NP-1 designation, a Ryan civil low wing trainer known as the NR-1 and similar to the Army's PT-22, and another Timm civil Aeromold low wing monoplane as the N2T-1.

Such was the need for trainers in the early World War II period that many Navy Piper Cubs were used under the designation NE-1.

A post war primary trainer tested by the Navy in 1947 was the clean cut Fairchild XNQ-1 but though tested by both services the type never progressed past prototype stage.

The later Navy trainer types are covered under the VT trainer series.

CURTISS N-9H (Mfr.) A standard Navy two place training seaplane of late world War I and the early 1920s, the N-9H in the photo was an uprated N-9 powered by the 150 horsepower Wright Hispano engine in place of the N-9's 100 horsepower Curtiss engine. N-9Hs differed in detail; about 300 airplanes were built, just a few by Curtiss and most by the Curtiss-owned Burgess Company of Marblehead, Ma. Some were also assembled from spare parts. The Hisso engine gave the 2750 pound gross weight N-9H a high speed of about 78 mph.

BOEING NB-1 (Mfr.) A total of 41 Boeing NB-1 two bay zero-stagger biplane trainers designed as landplanes convertible to seaplanes as shown in the photo were ordered by the Navy and delivered starting in December, 1924. Some were used as gunnery trainers. The NB-1 used Lawrence or Wright radial engines of 200 to 220 horsepower. NB-1 landplanes weighed about 2840 pounds and had a high speed of 100 mph. At the Navy's request another 30 trainers designated NB-2 were powered by 180 horsepower Wright Hispano water cooled engines. A large number of these NB- trainers were based at Pensacola, Fl. and served for several years. Some also went to the USMC. Five of these trainers were exported to Peru.

BOEING XN2B-1 (Mfr.) A Boeing Model 64 trainer prototype using a 200 horsepower Wright Whirlwind engine was proposed as a trainer replacement for the NB- aircraft, but was not accepted, and a further development, the Boeing Model 81, was built in 1927-8 as the XN2B-1 for a Navy trainer competition against the Curtiss XN2C-1 and the Keystone XNK-1 originally using the 127 horsepower Fairchild-Caminez engine. This engine gave the 2180 pound XN2B-1 trainer a top speed of 104 mph as a landplane. The engine installation proved unsatisfactory and a Wright Whirlwind engine of 165 horsepower was substituted.

BOEING XN2B-1 (Mfr.) Another view of the Boeing XN2B-1 shows the Fairchild-Caminez engine installed driving a two bladed propeller. A four blade propeller was also tested on the airplane. The powerplant had major problems, and the Boeing entry lost the competition to the Curtiss XN2C-1. There was no further production except for a civil Model 81A that went to the Boeing School of Aeronautics.

CURTISS XN2C-1 (Mfr.) A company photo of March, 1928 shows the Curtiss entry in the 1927-8 Navy trainer competition, the XN2C-1 Fledgling large double bay biplane with a 39 foot wingspan, one of three procured by the Navy and the competition winner. The airplane was a navalized development of the civil versions used by the Curtiss Wright Flying Service. The XN2C-1s tested by the Navy had a 200 horsepower Wright Whirlwind radial engine, weighed 2600 pounds loaded, and had a high speed of just over 100 mph.

CURTISS XN2C-1 (Court Commercial) A photo taken by a commercial house often used by Curtiss depicts an unusual view of one XN2C-1 prototype "thundering" past at very low altitude, probably doing at least 100 mph! Such action photos were a bit unusual in the early days; most company and naval pictures were "set" static ground shots around the aircraft.

CURTISS N2C-1 (H.Thorell) A production run of 31 N2C-1 models by Curtiss went to the Navy using the same 200 horsepower Wright Whirlwind and was a bit faster at about 108 mph and had a gross weight of 2830 pounds. Used mainly at Navy and USMC Reserve bases for primary training, the big awkward looking N2C-1 Fledgling (the civilian name) biplane served well into the 1930s as the last double bay wing type in the Navy.

CURTISS N2C-2 (H.Thorell) An additional 20 Fledglings designated N2C-2 because of the installation of an uprated 240 horsepower Wright R-760 Whirlwind engine were procured. Note the exhaust collector forward of the cylinders and the exhaust pipe extended well aft in the photo. Some N2C-2 trainers were modified with tricycle landing gear as radio controlled drone targets for Navy ships.

PIPER NE-1 (Mfr.) A 1942 Navy version of the civilian J3C-65 Piper Cub (meaning a J3 with a Continental 65 horsepower O-170 model engine) and Army L-4 Grasshopper is shown in a company flight photo. Cubs were used extensively in early wartime for primary training by both Navy and Marines, a total of 230 being acquired. The approximately 1200 pound gross weight two place tandem seat aircraft had a high speed of about 90 mph. Wingspan was 35 feet three inches. A few more, obtained from the Army, were labeled as NE-2 and corresponded to a later L-4 model.

HOWARD NH-1 (Mfr.) A 1941 Navy Anacostia photo shows the Navy and Marine advanced navigation and instrument trainer version of the civil Howard DGA-15P five place high wing monoplane (the DGA stood for "damn good airplane" and the P was for use of a Pratt and Whitney engine). A total of 205 NH-1s were produced , and pilot's feelings towards the type were mixed; some thought it was tough to handle. The 450 horsepower Wasp Junior engine powered the 4350 pound gross weight NH-1 to a high speed of about 200 mph.

NORTH AMERICAN NJ-1 (H.Thorell) The NJ-1 was a Navy version (NA-28), 40 of which were procured by the Army for the Navy under an order dated December 14, 1936, of the Army BT-9 basic trainer. A clean cut low wing fixed gear two seater, the major difference from a BT-9 was use of a more powerful 500 horsepower Pratt and Whitney Wasp engine. Intended for training use at NAS Pensacola, Fl., some NJ-1s were used as high level staff transports. The fuselage was a fabric covered steel tube structure.

NORTH AMERICAN NJ-1 (USN) The first of many North American low wing monoplane trainers for the Navy, the NJ-1 was at that time a thoroughly modern aircraft. When World War II started there were 39 NJ-1s stationed at NAS Pensacola. The Pratt and Whitney Wasp engine provided the 4440 pound gross weight aircraft with a top speed of about 170 mph. Wingspan was 42 feet.One aircraft was tested as an NJ-2 using a Ranger engine, but no more of this version were built.

KEYSTONE XNK-1 (Mfr.) A May, 1928 photo shows the Keystone entry in a Navy trainer competition of that year. The other aircraft were the Curtiss XN2C-1 and Boeing XN2B-1. After three prototypes were tested both as landplanes and single main and tip float seaplanes, a total of 16 production NK-1s were procured with revised tail surfaces. Powered by a Wright R-790 Whirlwind engine of 220 horsepower the landplane version of the NK-1 had a gross weight of 2660 pounds, a wingspan of 37 feet, and a maximum speed of about 110 mph.

MARTIN N2M-1 (USN) Shown in a Navy flight photo of June, 1926, the single N2M-1 was distinguished by large I struts between biplane wings. The Martin aircraft was competitive with the Boeing NB-1 and Naval Aircraft Factory N2N-1 trainer designs, but was unsuccessful in obtaining a production contract.

NAVAL AIRCRAFT FACTORY N3N-1 (H.Thorell) After a single XN3N-1 prototype was delivered to NAS Pensacola, Fl. for testing in 1936 and appropriate revisions made an initial production run of 85 N3N-1 trainers, convertable from landplane to seaplane in two hours, were started in delivery in mid-1936. A total of 95 additional trainers were produced through early 1938. Very early N3N-1s had 220 horsepower Wright Whirlwinds built up by the Navy; later aircraft were powered with 235 horsepower Wright engines. As a seaplane the N3N-1 weighed about 2850 pounds and had a high speed of about 115 mph and cruised at about 100 mph.

NAVAL AIRCRAFT FACTORY XN3N-2 (H.Thorell) The NAF produced a single XN3N-2 with new tail surfaces and a variable pitch propeller in 1936 also. Later modified to include a rather cumbersome looking cockpit hatch system as shown, the aircraft was in Navy service until just before US entry into World War II. Powered by a 240 horsepower Wright Whirlwind engine, the aircraft grossed out at 2650 pounds and had a landplane high speed of about 125 mph.

NAVAL AIRCRAFT FACTORY N3N-3 (USN) After modification of an N3N-1 into an XN3N-3 of 1939 with a new tail again and a simplified landing gear, production of the N3N-3 started in April of 1940 and lasted into early wartime January, 1942 with a total of 816 trainers being produced. The new landing gear is shown in this Navy flight photo (*above*). The N3N-3 (*below*) was powered by a Wright R-760 of 235 horsepower and as a landplane had a gross weight of about 2800 pounds and a 125 mph high speed.

NAVAL AIRCRAFT FACTORY N3N-3 (UNK.) Shown in single main float seaplane form, the popular N3N-3 trainer turned out to be the last biplane in either Navy or Army service, being used as a flight indoctrination trainer for naval cadets at Annapolis as late as 1960. The N3N-3 was delivered in quantity to both NAS Pensacola and NAS Corpus Christi. These aircraft were used at the same time as the Stearman N2S- biplane trainers.

NAVAL AIRCRAFT FACTORY XN5N-1 (USN) The Naval Aircraft Factory in Philadelphia designed and built a new low wing monoplane primary trainer prototype meant to supersede the N3N-3 biplane as shown in the Navy photo of April 18, 1941. Only one aircraft was constructed. Powered by a Wright R-760 engine of 350 horsepower the XN5N-1 trainer had a gross weight of about 3300 pounds, a top speed of approximately 135 mph, and a wingspan of 42 feet. The plane was tested with both open cockpits as shown and with a closed canopy. No production resulted from this effort.

SPARTAN NP-1 (P.Bowers) The initial aircraft of a Navy order of mid-1940 for a navalized Spartan C-3 civil sport biplane is shown in a 1941 photo. Used as a primary trainer at Naval Reserve bases starting that year, it was powered with a 220 horsepower Lycoming R-680 radial engine, had a 3000 pound gross weight, and a maximum speed of about 105 mph.

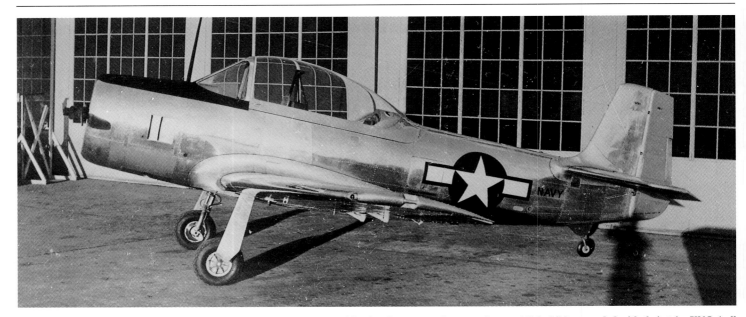

FAIRCHILD XNQ-1 (Mfr.) Late in World War II the Navy set out a specification for a new primary trainer and Fairchild responded with their trim XNQ-1 all metal low wing monoplane with retractable landing gear. Mockup inspection took place in September, 1945 and first flight in October, 1946. A second aircraft was ready for flight test in February, 1947. These planes were also considered for production as USAF T-31 trainers. Powered by a 320 horsepower Lycoming R-680 engine high speed was about 175 mph. One of the aircraft was also tested with a new horizontally opposed 350 horsepower Lycoming GSO-580 engine. Lack of funding caused the demise of the program and no production took place.

RYAN NR-1 (USN) Flight photo of the Navy version of the Ryan two seat civil model ST-3 and Army PT-21 labeled an NR-1. Ordered in the summer of 1940, a total of 74 NR-1 production versions were supplied to the Navy for primary training use after testing of a single prototype. Powered by a 132 horsepower Kinner R-440 radial engine, the NR-1 had a wingspan of 30 feet one inch, a gross weight of 1825 pounds, and had a high speed of about 120 mph. Service ceiling was about 15500 feet.

STEARMAN NS-1 (Mfr.) The initial Navy version of the famous Stearman Kaydet primary training biplane, company Model 73, a development of the original Model 70 tested by the Navy in March of 1934. The production NS-1s were delivered in two batches, 41 and 20 totaling 61 aircraft in 1935-6. The Navy had a surplus of Wright J-5/R-790-8 Whirlwind engines of 225 horsepower and specified use of these engines in the trainer. This two place biplane had a wingspan of 32 feet two inches, a gross weight of 2700 pounds, and a maximum speed of 118 mph.

STEARMAN N2S-1 (Mfr.) The next Navy version of the Stearman trainer was the Model A75N1 known in the Navy as the N2S-1 which was similar to the Army PT-17 in using the Continental R-670 radial engine of 220 horsepower. A total of 250 trainers of this type were delivered between September, 1940 and February, 1941. The 2710 pound normal gross weight biplane had a top speed of 125 mph and a service ceiling of 13400 feet. In addition 125 N2S-2 variants using the Lycoming R-680 engine with the same performance were delivered between April and October of 1941; these were the equivalent of Army PT-13As.

STEARMAN N2S-3 (R.Dean) The major production version of the Stearman B75N1 primary trainer for the USN, again using the Continental R-670 engine. A total of 1875 N2S-3s in five batches were delivered from February, 1941 through June of 1943, and were little different from the N2S-1s. A very few had cockpit canopies. Specifications pretty much duplicated those of the -1 version.

STEARMAN N2S-5 (Mfr.) After production of 577 N2S-4s which were nearly the same as the N2S-1 with the Continental engine, Stearman standardized on the N2S-5 which was the same as the Army PT-13D and identified as company Model E75. These -5 Kaydets were delivered from June, 1943 to February, 1945 with 873 N2S-5s going to the Navy. These were all silver colored when delivered. The clean contours of an N2S-5 Kaydet are nicely shown in this manufacturer's photo of July 5, 1944. The 220 horsepower Lycoming R-680 powered the 2720 pound Stearman to a maximum speed of 124 mph.

NEW STANDARD NT-1 (H.Thorell) A 1930 navalized biplane trainer based on the civil New Standard D-29, six of which were purchased for Navy evaluation against the other primary trainer candidates with the hope of finding the lightest and lowest possible cost suitable model. There was no further production. The engine was a Kinner K-5 five cylinder radial type of about 110 horsepower. Gross weight was only 1800 pounds and high speed almost 100 mph.

TIMM N2T-1 (F.Dean) The photo shows a post World War II civil conversion of a Timm N2T-1 two seat naval primary trainer originally known as a civil PT-175K. Mid-wartime deliveries were made of 262 N2T-1s to the Navy and Marine Corps. Its claim to fame was use of Timm Aeromold construction, plywood bonded with plastic. The engine was a 220 horsepower Continental R-670 radial giving the 2700 pound gross weight low wing monoplane a maximum speed of about 140 mph.

TIMM N2T-1 (USN) The Navy photo shows a view of the N2T-1 in service markings. Noteable is the use of considerable wing dihedral. The Timm Tutor was completely conventional in arrangement but was distinguished by the Aeromold bonded plywood construction with the plus of saving on aluminum, once thought to be a wartime supply problem.

.TIMM XN2T-2 (USN) A Timm Tutor primary trainer was modified to use a horizontally opposed engine, probably a Lycoming model, and labeled an XN2T-2. Shown in a Navy photo of November 22, 1943, this version of the Timm Aircraft Company's trainer was not put into production.

CONSOLIDATED NY-1 (UNK.) A Navy primary trainer version of the Army's PT-1 with, however, an aircooled radial engine, the nine cylinder Wright J-4 of 200 horsepower, the NY-1 of 1926 was the first Navy airplane from Consolidated, and as the photo shows was convertable to a single main and tip float seaplane. Another change from the PT-1 was the larger tail surfaces. Seventy-six NY-1s were purchased. Though there was some overheating of the air cooled radial in the warm Pensacola climate, the NY-1 was an effective trainer. As a gunnery trainer the plane was known as the NY-1A. Span was 34 feet six inches, seaplane gross weight 2750 pounds, and maximum speed about 100 mph as a seaplane and slightly more as a landplane.

CONSOLIDATED NY-2 (H. Thorell) Shown as a landplane running up in the photo is an NY-2 Navy Husky of 1928 of which that service bought not less than 133 aircraft. A counterpart of the Army's PT-3, the NY-2 was an updated NY-1 with greater wingspan of 40 feet but the same engine. Like the NY-1 it was convertable to a seaplane and was used as a trainer and an observation plane. High speed was about 95 mph and gross weight as a seaplane was just over 2900 pounds.

CONSOLIDATED NY-2A (H.Thorell) The NY-2A designation was used to denote an armed version of the NY-2 used as a gunnery trainer, however the extent of the armament used is unclear, and it is not shown in this photo of the landplane variant. Simplicity of design was the byword; note the single piece rectangular windshields. An NY-2 was used by Jimmy Doolittle to demonstrate blind flying capability.

CONSOLIDATED NY-2A (H.Thorell) A moored NY-2A with cockpit and engine covers installed shows the seaplane configuration of the trainer. The airplane type lasted well into the 1930s; some went to the Naval Reserve. Ailerons were located on both wings with a drive strut in between.

CONSOLIDATED NY-3 (H.Thorell) The NY-3 was similar to predecessor variants except for an uprated Wright R-760 Whirlwind engine of 240 horsepower. About 20 of these were procured and went to the Reserves. Gross weight of this landplane version was 2600 pounds with high speed about 105 mph. These antiquated looking biplanes served in the Navy through the 1930s.

CONSOLIDATED XN2Y-1 (Mfr.) A completely new Consolidated Fleet primary trainer was the XN2Y-1 of 1929 using the Warner Scarab engine of 110 horse-power. This airplane was simply a naval version of the civil Fleet 1 biplane. With gross weight of just under 1600 pounds the trim little biplane with ailerons only on lower wings had a high speed of about 105 mph. No production was forthcoming with the Scarab engine, but a slightly modified N2Y-1 using a five cylinder Kinner engine was used by the service. One of these N2Y-1 Fleets was equipped with an airship hook on the upper wing and in 1931 was used as a trainer for airship pilots of the Akron and Los Angeles.

CONSOLIDATED XN3Y-1 (USN) One NY-2 was updated in 1929 with a new engine version of the Wright Whirlwind radial and minor wing revisions. The design looked primative, even for 1929. There was no N3Y-1 production.

CONSOLIDATED N4Y-1 (H.Thorell) A fine photo of the clean cut Coast Guard version of the civil Model 21 used by the Army as the PT-11. A new trainer designed with the idea of providing both primary and more advanced training in 1931, only the one aircraft shown was purchased by the Coast Guard using the Wright Whirlwind engine of 165 horsepower, and was used right up until World War II. Three other N4Y-1s were procured by the Navy using 200 horsepower Lycoming engines. Span was 31 feet six inches and gross weight around 2500 pounds. High speed was about 120 mph.

NAVAL AIRCRAFT FACTORY TG-3 (USN) One of five prototype seaplane double bay biplanes built by the NAF in 1924 as gunnery trainers is shown at the Philadelphia factory. This single main and tip float seaplane could be tested in the adjacent Delaware River. The five aircraft were designated TG-1 through TG-5 with the TG-3 shown powered with an Aeromarine T-6 engine of 200 horsepower. Gross weight was about a ton and a half and high speed almost touched 100 mph.

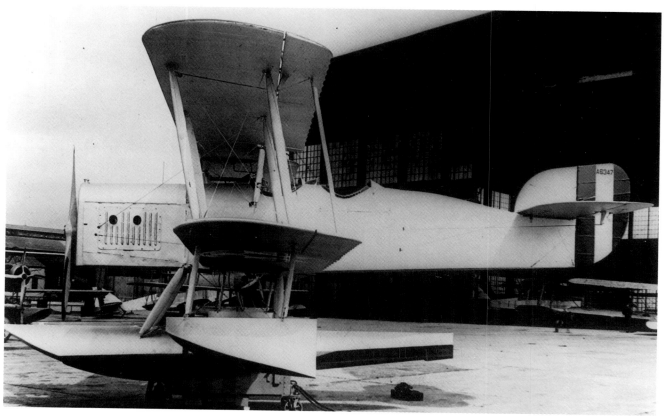

NAVAL AIRCRAFT FACTORY TG-4 (USN) Another of the prototype NAF trainer seaplanes , a TG-4 with the same Aeromarine engine and similar characteristics, is shown in this Navy photo. Other planes of the TG- series were powered by either Liberty or Wright E engines. There was no production resulting from these NAF prototypes.

NAVY OBSERVATION AIRPLANES

The Navy VO observation class of airplanes started right after World War I in the early 1920s. In 1921 the Navy Bureau of Aeronautics came into being and airplane development programs became more organized. Observation planes were used for gunfire spotting and were often designed as single or twin float seaplanes, convertable to landplanes for Marine use, and some were designed as amphibians. Observation planes were for years a stalwart part of the fleet. They were based not only on aircraft carriers but on major combat ships such as battleships and cruisers.

Certainly the most successful of the early observation aircraft was the Vought UO- series designed around the Wright Whirlwind engine. The UO-1 of 1922 was convertable, capable of use either on the carrier Langley or off capital ship catapults. The UO- type served with the fleet for several years.

In 1923 and later the Marines used improved DeHavilland DH-4s built by Boeing as O2B-1s for ground attack and observation duties with the Expeditionary Forces. At this time several naval observation types appeared, the prototypes-only Huff Daland HO-1 version of their Hisso powered HN-1 trainers, the three place Curtiss D-12 powered Martin MO-1 metal monoplane, convertable for wheels or floats, with a production order but not really successful, and the Naval Aircraft Factory NO-1 convertable biplane of which several were built using the D-12 engine.

Another successful naval observation type starting in 1925 was Grover Loening's outstanding OL- large biplane amphibian series, many versions of which lasted into the 1930s powered with various engines but ending with a Wasp air cooled radial. The OL- designs, -1 to -9, were used both in fleet operations and for various naval exploration expeditions.

In 1927 two observation types appeared, the Curtiss OC-2 Wasp powered two seater landplane for the Marines, originally designated as fighters, and the Vought O2U- Corsair, a significant advance and one of the most famous of pre-World War II naval planes put out in several variants, including landplane, single float seaplane, and amphibian.

In 1929-30 the Navy bought a couple of Army O-2 landplane observation types for the Marines as OD-1s, and later some Wasp powered Curtiss O2C- Helldiver two place landplanes originally labeled as F8C-5 fighters for that service. Some of these Helldivers became VIP transports.

In 1930 Vought started producing a follow-on Corsair model, the O3U- variants put out in substantial numbers. It operated with the fleet for several years both as a battleship and cruiser catapult aircraft and as a carrier plane in utility units. The O3U-s were also used as landplanes, single float seaplanes, and amphibians.

The year 1931 saw two prototype observation planes tested by the Navy, the Loening XO2L-1 and the Vought XO4U-1. both types developments of the OL-8/-9 amphibians and the O3U- respectively, utilized all metal construction except for wing covering material. Neither model reached production and service use however. About this time also Pitcairn XOP-1/-2 autogiros were evaluated for observation duty but were rejected. An additional plane evaluated in 1931 was the Keystone XOK-1 light observation biplane for catapult use; there was no production.

A year later Berliner Joyce developed a successful Wasp powered light observation plane for light cruiser catapult operations as a seaplane and later a landplane for the use of reserves. This was the OJ-2, a conventional two place biplane.

In 1934 there was a Navy competition for a Corsair replacement as a cruiser and battleship catapult plane originally required to be an amphibian, but that requirement was later dropped. Three biplane aircraft prototypes were built for the competition, the Curtiss XO3C-1, the Douglas XO2D-1, and the Vought XO5U-1. The Curtiss machine won out and was subsequently developed into the famous SOC- Seagull catapult aircraft.

In 1936 the last pre-World War II observation plane design was procured for the US Coast Guard, the Viking OO-1, a small flying boat of French origin.

During early World War II the Navy procured a batch of Stinson Army L-5 liaison planes, variants of the civil model 105 type, for the Marine Corps and designated it the OY-1.

After World War II there were three Navy airplane types coming under the observation class, the Cessna OE- version of the Army/Air Force L-19 Birddog series in 1955, the unusual 1957 Pennsylvania Aircraft Syndicate XOZ-1 convertaplane on floats with an unpowered rotor, and the North American Rockwell OV-10 twin engined Bronco of 1968 used for light armed reconnaissance and forward air control by the Marine Corps.

BOEING DEHAVILLAND O2B-1 (USN) A fine Navy flight photo of a Boeing-modernized DH-4 airplane called a DH-4M, later the O2B-1. From 1923-5 a total of 187 DH-4Bs were rebuilt with steel tube fuselages, and 30 of these were provided in 1925 to the USMC as observation types. These aircraft were used both in China and Nicaragua by the Marines. Wingspan was 42 feet five inches; armament consisted of four machine guns and a 400 pound bomb load. Liberty engine powered, the 400 horsepower O2B-1s weighed 4600 pounds gross and had a maximum speed of 118 mph.

CURTISS OC-1 (C. Mandrake) A Marine Corps OC-1 of late 1927 to early 1928 is one of many service versions of the Curtiss Falcon with this version used to replace DH-4s. Originally designated F8C-1s, four of these planes were utilized by the Marines in Nicaragua fighting. Like the DH-4Ms they had two fixed and two flexible guns and could carry light bombs. Unlike Army production Falcons these planes were powered by 410 horsepower Pratt and Whitney Wasp engines. This two seater observation type weighed a little over 3900 pounds loaded, had a wingspan of 38 feet, and a high speed of about 135 mph.

CURTISS OC-2 (MFR.) A Curtiss photo of mid-April, 1928 shows an OC-2 Falcon for the Navy right out of the factory with the mounting ring for twin Lewis guns in the rear cockpit. One of 21 observation planes originally designated as F8C-3 two seat fighters, this designation was quickly found inappropriate and changed to an OC-2 observation type for the two Marine observation squadrons.

CURTISS OC-2 (USAF) A photo taken on April 27, 1930 shows one of the Marine Corps OC-2 Falcons without the rear gun mount installed and USMC insignia alongside the cockpits. Sweepback of outer panels of the upper wing is readily apparent. An engine starter crank is secured alongside the forward fuselage. Powered with a 430 horsepower Pratt and Whitney Wasp engine, the approximately two ton gross weight OC-2 could achieve a high speed of about 140 mph.

CURTISS XOC-3 (Mfr.) An October, 1930 manufacturer's photo shows one of the OC-2 Falcons converted into an XOC-3 variant because of the installation of an experimental Curtiss H-1640 "Hex" engine rated at 600 horsepower, quite a power boost for the Falcon. The installation was also tried in an Army XO-18 Falcon but neither installation was successful. The engine was a failure and no production resulted. The plane had a gross weight of 4300 pounds and the extra installed power increased top speed to about 145 mph.

CURTISS O2C-1 (P.Bowers) A Curtiss Helldiver from the Naval Reserve Air Base in Seattle, Wa. is shown in the mid-1930s. Essentially a new and smaller design compared to the Falcon, though with the same basic engine, the O2C-1 was originally procured as a two seat F8C-5 fighter, but this was shortly changed to the observation designation. Capable of serving as a 500 pound bomb dive bomber, this Helldiver had two fixed guns in the upper wings and a flexible rear cockpit weapon. Power came from a ring cowled Pratt and Whitney 450 horsepower Wasp engine. Ninety-three Helldivers went to the Marines starting in late 1930.

CURTISS O2C-1 (Mfr.) There were several variants of the O2C-1 Helldiver including this one with enclosed cockpits for VIP use of such officers as Cdr. Dillon (the photo legend says). All armament was discarded in this naval air taxi.

CURTISS O2C-1 (H.Thorell) Later on O2C-1 Helldiver airplane number 8945 was modified as another version of a VIP transport, back to open cockpits, though a headrest was installed for the rear cockpit. A noteable addition was a pair of pretty snappy wheel pants. This airplane was apparently assigned to Naval Reserve Air Base, Brooklyn, NY.

CURTISS O2C-2 (Mfr.) A September, 1931 Curtiss photo shows one of three O2C-2 special Helldivers with partially enclosed cockpits and wheel pants. Three of these models were built using Wright R-1820 nine cylinder Cyclone E engines of 575 horsepower. Note the Helldiver was armed and equipped with a tail hook, and it did serve for awhile on a carrier. The additional power of the Cyclone engine gave these approximately 4500 pound gross weight versions a top speed of about 175 mph. This plane was later revised into an XS3C-1 scout type which crashed in 1932.

CURTISS XO3C-1 (Mfr.) The prototype that won a new Navy observation aircraft competition of the early 1930s, the Curtiss XO3C-1 in modified form turned into the SOC-Seagull which was to become a highly successful production airplane both in landplane and seaplane form. The XO3C-1 beat out competitor Douglas XO2D-1 and Vought XO5U-1 prototypes. The amphibian wheels retracted into the main float and the cockpits were open, these features being peculiar to only the prototype.

CURTISS XO3C-1 (Mfr.) The XO3C-1 prototype sits on the tarmac amid the melting snow of March, 1934, two days before its first flight. Showing are the amphibious gear, a tail hook, upper wing flaps, and lower wing ailerons. Powered by a Pratt and Whitney R-1340 Wasp engine of 550 horsepower, the 5200 pound XO3C-1 had a high speed of about 160 mph. It later became an XSOC-1 and was wiped out in a 1941 crash.

DOUGLAS OD-1 (USN) A rather rare bird is displayed in this Navy photo of March 6, 1929. One of only two Douglas O-2C Army types redesignated an OD-1 for the US Marines use on the west coast; as shown in the photo these planes were apparently used mainly for utility duties rather than as observation planes. Powered with the Liberty 12 engine of 400 horsepower and with a wingspan of 39 feet, eight inches, the approximately 4500 pound gross weight plane had a maximum speed of about 130 mph.

DOUGLAS XO2D-1 (Mfr.) One of the two losing prototypes to the Curtiss XO3C-1 in the 1933-35 competition for a shipboard catapult observation amphibian, the Douglas XO2D-1 was a slick looking biplane with a canopy over tandem cockpits. Delivered in the spring of 1934, it was also tested as a pure seaplane. Like the Curtiss entry it was powered by a 550 horsepower Pratt and Whitney Wasp engine which gave the 5100 pound amphibian a high speed of about 160 mph. As a competition loser there was no production, and the aircraft became just another obscure type.

CESSNA OE-2/O-1C (R.Besecker) A mid-1950s update of the Army L-19 and Navy and Marine OE-1 Birddog, the OE-2 spotter and forward air control type was designed specifically for the Marine Corps. Two wing store stations could carry light armament loads like rockets, and the aircraft was equipped with armor and self sealing fuel tanks. Only 27 were built, however. Powered by a 260 horsepower Continental horizontally opposed engine, the 2650 pound gross weight OE-2 (changed to an O-1C designation in the 1962 redesignation plan) had a high speed of over 180 mph.

HUFF DALAND HO-1 (USN) An unarmed Navy observation plane of 1923 is shown in the photo; typical of the times it was convertible to a twin float seaplane. Alnost identical to six HN-1 and HN-2 trainer versions, the three HO-1s were the only observation models made by Huff Daland. Note the unusual location of the national insignia inboard on the wings. Powered by a 180 horsepower Wright Hispano E engine the HO-1 landplane version had a gross weight of 2300 pounds and a high speed of about 100 mph. Wingspan was 33 feet.

BERLINER JOYCE XOJ-1 (H.Thorell) One of two 1931 light weight prototypes (the other was a Keystone XOK-1) in a competition for a catapult land/seaplane observation type for Navy light cruisers, the XOJ-1 was distinguished by unusual full span flaps on both biplane wings, shown deflected in the photo, and "park bench" type ailerons separate from and above the upper wing panels. This arrangement for low landing speed capability was not pursued on production OJ-2s. Powered by a 400 horsepower Pratt and Whitney Wasp Junior engine, the 3200 pound gross weight plane could make a top speed of about 150 mph.

BERRLINER JOYCE OJ-2 (E. Sommerich) A fine photo of the lightweight Berliner Joyce observation plane during its later life in the Naval Reserves shows the type reverted to a conventional ailerons-on-both-wings configuration. There were 39 production OJ-2s produced and they started going to sea with Navy light cruisers in 1933. After replacement by Curtiss SOC-1s these planes were used in the reserves through the 1930s.

BERLINER JOYCE OJ-2 (H. Thorell) A view of this OJ-2 shows a completely conventional biplane with provision for carrying stores under each lower wing. One fixed forward and one flexible machine gun could be carried. Production was split between sea-based cruiser catapult seaplanes and Naval Reserve landplanes. A few stayed at naval bases until the late 1930s. Landplane high speed was about 150 mph. Wings spanned 33 feet eight inches.

BERLINER JOYCE OJ-2 (USN) The OJ-2 as a single main and tip float seaplane is shown in this Navy photo of Squadron VS-5B plane number five. The MEM on the handling dolly reminds us it was apparently the property of light cruiser USS Memphis. As a seaplane the 400 horsepower Pratt and Whitney Wasp Junior powered the 3850 pound gross weight OJ-2 to a top speed of about 145 mph. The OJ-2s were replaced by Curtiss SOC-Seagulls.

BERLINER JOYCE XOJ-3 (H.Thorell) A photo of the NAS Anacostia-based XOJ-3 landplane, a 1934 modification of an OJ-2 in an attempt at modernization. Major changes included a new three blade propeller, a long chord engine cowl, metal skinned forward fuselage, and an interesting cockpit canopy arrangement. After a 1935 crash the XOJ-3 was rebuilt pretty much back as an OJ-2 and served in the Reserves until 1940.

LOENING OL-1 (USN) The initial Loening amphibian for the Navy appeared in mid-1925 when two OL-1 three place aircraft were procured, distinguished by a main seaplane float integrated into the fuselage design and wheeled landing gear flipping over into wheel recesses in the float. The double bay wings were interchangeable with those of the DH-4. The OL-1 incorporated a Packard IA-1500 engine of 440 horsepower driving a four blade propeller whose diameter was limited by float tip clearance. The OL-1s could be catapulted from battleships. They were large; span was 45 feet, gross weight 5200 pounds, and they had a high speed of 125 mph.

LOENING OL-2 (USN) This Navy photo of June 10, 1925 shows one of five OL-2 amphibians taxiing out for a takeoff. Two place instead of three, and powered by a 400 horsepower Liberty engine, these OL-2s were used in Arctic exploration and later by the Marine Corps. With such a maze of struts it hardly seems possible this 5000 pound gross weight machine could attain a 120 mph top speed!

LOENING OL-3 (USN) With tail skid up on a dolly, this OL-3 provides a good view of the wing and landing gear arrangement of the Loening amphibians. Basically a three place OL-2 using the inverted Packard engine uprated to 475 horsepower, three or four of these Loenings were used for Alaskan area mapping in 1926. Slightly heavier than the earlier models, the OL-3s had pretty much the same performance.

LOENING OL-4 (USN) The four to six Loening amphibians of 1926 reverted back to use of the 400 horsepower Liberty engine but the engine drove a three bladed propeller; the configuration was otherwise similar to earlier models. Gross weight was up to about 5450 pounds and maximum speed down to around 117 mph. Three Liberty powered OL-5 versions pioneered US Coast Guard aviation starting in 1926. Each cost $32,710.

LOENING XOL-7 (USN) After 27 OL-6 Loenings had been manufactured, these three Packard-powered models came along. The Navy ordered one modified with a new wing cellule of the single bay type with a manual folding capability added for carrier use. Gross weight was considerably higher. Tested in 1927, it was found unsuitable and no production was undertaken.

LOENING OL-8 (USN) A new look was imparted to the Loening amphibian with use of the Pratt and Whitney Wasp radial air cooled engine of 450 horsepower after tests of a modified OL-6. The lighter engine reduced gross weight to little over 4800 pounds with no loss in speed. This December, 1928 photo shows a VIP command transport version (note the two stars in two places!) as one of 20 aircraft procured in 1927-28.

LOENING OL-8A (USN) An additional 20 amphibians using the Wasp engine were obtained by the Navy in 1929 and designated OL-8A; whether the A designated bomb-carrying capability or carrier hook equipment or both is not clear. The airplane shown is one of those used during another mapping project of Alaska in 1929. Some of these amphibians lasted well into the 1930s.

LOENING OL-9 (USN) Now actually a Keystone Loening amphibian since Keystone took over Grover Loening's company and produced 26 armed Wasp-powered versions in 1931-32 somewhat heavier than the OL-8s, but with little performance difference. These amphibians were replaced in the mid-1930s by the Grumman JF- Duck using a modernized but very similar configuration.

KEYSTONE LOENING XO2L-1 (USN) These Navy photos (*center, right*) of September 16, 1931 show the next effort of Keystone to keep the Navy amphibian line going with a couple of new prototypes, the XO2L-1 being the first; these completely modernized designs in the same configuration. The aircraft was ahead of the similar Grumman JF-1 by about three years. Note the unusual vertical tail surfaces added and the carrier hook.

KEYSTONE LOENING XO2L-1 (USN) An interesting Navy photo of October 15, 1931 shows the XO2L-1 prototype sitting in rough water apparently without the Wasp engine fired up. The Pratt and Whitney R-1340 of 450 horsepower gave the 4050 pound gross weight aircraft a high speed of about 130 mph. Only two prototypes were built; both were tested aboard ship.

KEYSTONE LOENING XO2L-2 (USN) The second of the Keystone prototypes is shown in a February 10, 1932 photo with the major configuration difference from the later Grumman JF- series being open cockpits and a different gear retraction method. The only visible difference from the XO2L-1 was a revised vertical tail, but an uprated Wasp engine of 550 horsepower was installed adding ten mph to high speed even though gross weight was several hundred pounds greater at 4800 pounds.

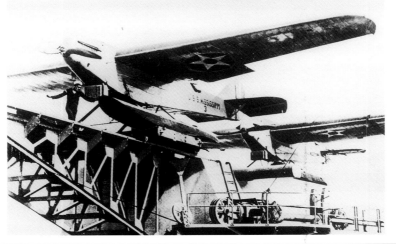

MARTIN MO-1 (Mfr.) A three place heavy observation design by Martin in either land or seaplane form is shown aboard a catapult on the USS Mississippi. Based on all metal structure like German Junkers designs, the MO-1 monoplane was ordered to the extent of 36 aircraft in 1923 and delivered through that year. Unsuccessful at sea, they ended up operating as landplanes powered by a 350 horsepower Curtiss D-12 engine. Wingspan was 53 feet, gross weight about 4600 pounds, and maximum speed around 105 mph.

NAVAL AIRCRAFT FACTORY NO-1 (USN) One of three three place heavy observation planes by the NAF from a Navy design is shown in a March, 1924 photo. Another three similar aircraft were built by Martin as M2O-1s. No more production took place. Powered by a Curtiss D-12 engine of 350 horsepower, the NO-1 seaplane had a maximum speed of about 105 mph and a gross weight of approximately 4200 pounds. A maze of wing interplane struts characterized this biplane design.

NAVAL AIRCRAFT FACTORY NO-1 (Mfr.) Another NO-1 Navy design, this one built by Lowe, Willard, and Fowler Engineering company under Navy contract 57340. The aft cockpit for the third man is shown equipped with a Lewis gun on a Scarff ring mount. No picture has been seen of a landplane version. One NO-1 was modified to an NO-2 by installation of a Packard engine.

VIKING OO-1 (H.Thorell) One of five "Lollypop" 43 foot four inch span biplane flying boats built for the US Coast Guard in 1936 is shown. These aircraft were built by the American Viking company based on a French Schrenck design, and could be found at east and gulf coast USCG bases. Powered by a 240 horsepower Wright Whirlwind engine in a pusher installation, the 3450 pound gross weight "Lollypop" had a high speed of just over 100 mph.

VOUGHT UO-1 (Wright) A landplane version of the 1922 Vought UO-1 two seat two bay wing biplane of which about 140 aircraft were made is shown. The convertable land and seaplane design was an outgrowth of the earlier VE-7 and VE-9 models, in fact wings were the same. Major changes took place in fuselage re-contouring, a new vertical tail, and most importantly installation of a new Lawrence or Wright radial engine of 200 horsepower. Note the outside metal surfaces of the fuel tanks just outboard of the front cockpit.

VOUGHT UO-1 (USN) In an unusual photograph a UO-1 two seat two bay biplane seaplane from the battleship Tennessee trails in echelon just behind its predecessor VE-7 seaplane from the cruiser Milwaukee. The contrast in fuselage contours is very apparent, but the new radial engine is masked by the upper wing. How the sailor in the rear cockpit is keeping his hat on is a question! The national insignia was placed to avoid aileron and float areas. UO-1s were aboard battleship catapults in the 1920s.

VOUGHT UO-1 (Wright) A UO-1 being recovered aboard the cruiser Richmond shows all the then-new features of the aircraft, an extremely successful design. Wingspan was 34 feet four inches and gross weight 2800 pounds. The 220 horsepower Wright Whirlwind in later aircraft powered this light observation seaplane to a maximum speed of about 120 mph. In 1925 the Coast Guard borrowed a UO-1 from the Navy and operated it very successfully from a base at Gloucester, Ma.

VOUGHT UO-4 (Mfr.) One of two UO-1 aircraft modified in late 1926 for the US Coast Guard is shown in a manufacturer's photo. These two seaplanes, designated UO-4s, along with three Loening OL-5 amphibians, formed the foundation on which US Coast Guard aviation rested. The two Voughts lasted well into the 1930s. Powered by a Wright R-790 Whirlwind radial engine of 220 horsepower, the 2900 pound UO-4 seaplane had a high speed of just under 120 mph.

VOUGHT O2U-1 (Mfr.) A very successful airplane is shown at the Vought plant in 1927; the first Navy Corsair, labeled an O2U-1 observation type. A follow-on to the earlier VE-7/VE-9 and UO-1 series, the Corsair was a trim single bay biplane with a Pratt and Whitney Wasp engine, ailerons on both wings, and fuel tanks built into the forward fuselage sides. The first of two prototypes flew in late 1926. The aircraft is shown prior to Navy squadron assignment. The Corsairs served faithfully aboard the catapults of major combat ships.

VOUGHT O2U-1 (H. Thorell) Counting the two prototypes Vought built 132 O2U-1 Corsairs. Some went to the Marine Corps in 1927-28; others went aboard Navy ships, and some were used as command transports for VIPs. This Corsair had an oversized rear cockpit windshield, a ring cowl around its Wasp engine, and panted wheels. The fuselage structure was steel tubing; wing structure was all wood.

VOUGHT O2U-1 (Mfr.) A Vought photo of an O2U-1 intended as a command plane done up with a proper paint job along with ring cowl and wheel pants. The sweep of the upper wing is well shown here. For war use the shipboard O2U-1s had a single .30 caliber machine gun in the upper wing, a ring for a Lewis gun in the rear cockpit, and racks for light bombs under the lower wings.

VOUGHT O2U-1 (USN) A famous photo of a Corsair seaplane from the cruiser Raleigh. The item above the center of the upper wing is a small wind-driven propeller powering an electrical generator for the aircraft radio. The 450 horsepower Wasp engine gave the 3900 pound gross weight Corsair seaplane a maximum speed of about 145 mph. Wingspan was 34 feet six inches.

VOUGHT O2U-1 (Mfr.) A manufacturer's photo of the fourth production O2U-1 Corsair shows it is being prepared as a slicked up command aircraft. High level naval air officers were eager to use this new Corsair for transport. The captain's or admiral's placard with appropriate stars could be slipped into the fuselage pocket alongside the rear cockpit.

VOUGHT O2U-2 (Mfr.) A view of a new O2U-2 Corsair with a rear cockpit headrest instead of a flexible gun installation and admiral's stars on a badge alongside the cockpit indicating VIP transport assignment. The 450 horsepower Wasp radial engine gave a 150 mph high speed capability to the 3635 pound gross weight Corsair. The side fuel tank filler can be seen just below the forward cabane struts.

VOUGHT O2U-2 (H.Thorell) With a rooster (Crowing Cock) insignia on its fuselage a Vought O2U-2 seaplane Corsair based on the carrier Saratoga sits at a lakefront dock. The 1928 O2U-2 differed only slightly from its predecessor in wing span (18 inches greater), dihedral, and a revised upper wing central cutout. Used with the fleet in both landplane and seaplane versions, the Corsair shown weighed about 3900 pounds and could hit a top speed of just over 145 mph.

VOUGHT O2U-2 (Mfr.) Some of the 37 O2U-2 Corsairs were fitted out as amphibians as shown in the photo. Main wheels were attached to the central float. Since the normal tail skid could not be utilized a skid or wheel was fitted at the rear of that float. Additional struts were placed between this aft point and the fuselage as seen here. A cut back section in the center of the upper wing can also be seen, possibly to provide better visibility upward from the front cockpit.

VOUGHT O2U-3 (Mfr.) Little was changed in the O2U-3 Corsair (*above*) models of 1927-30. A slight change in wing dihedral and revised tail surfaces were the major items; 80 aircraft were built for the US Navy and a few were exported, the example shown (*right*) believed to be headed for the Dominican Republic shortly after this manufacturer's photo of September 7, 1933 was taken. Wing span was 36 feet, landplane gross weight just under two tons, and the 450 horsepower Wasp engine gave the O2U-3 a high speed of about 140 mph.

VOUGHT O2U-4 (H.Thorell) Lean angular lines are emphasized in this view of one of 42 O2U-4 Corsairs built in 1929-30 (*below*). Very minor changes were made from the O2U-3. A rear gun ring could be substituted for the rear cockpit headrest. The leading edge cut-back in the center section is well shown here; not all Corsairs incorporated this arrangement. Weights and performance were similar to the preceding model.

VOUGHT O2U-4 (H.Thorell) A beached seaplane version of an O2U-4 from Pensacola is shown. Note this O2U-4 does not have the upper wing leading edge cutback. The rear cockpit is set up for a gun mounting ring, and the doors of a fuselage storage compartment are open. Differences between landplane and seaplane versions of the Corsair in terms of weight and performance were small.

VOUGHT O3U-1 (H.Thorell) An attractive photo of the next Corsair model shows little difference from the O2U-4 aircraft. Major changes were the lower wing sweep and span matched those characteristics of the upper wing. These Corsairs appeared in 1930-31 and were built after Vought had joined up with the United Aircraft Corporation. The Navy purchased 87 of this Corsair version; they loved the Corsair!

VOUGHT O3U-1 (Mfr.) Operating a Corsair seaplane from a battleship or cruiser was an art requiring careful handling and considerable skill as can be seen from this photo of an O3U-1 without power alongside a ship. The hoist, tie lines, and fender poles are all shown. The ship does not appear to be under way. Retrieving planes while under way was a real art!

VOUGHT O3U-1 (H.Thorell) Some O3U-1 Corsairs were converted in early 1931 with the new Grumman-type float making the aircraft into an amphibian as shown here. Main wheels retracted into the central float just like the many subsequent Grumman installations would in future years. Note also the tail wheel and local float-to-fuselage connection, possibly for steering; a structural support seems unlikely.

VOUGHT O3U-2 (Mfr.) Another new version of the Vought Corsair intended for carriers, shown in a manufacturer's photo, was the heavier and more powerful O3U-2 of 1931-32 equipped with a 600 horsepower Pratt and Whitney Hornet engine. Another change was use of a cross-axle landing gear and a carrier tail hook. A total of 29 O3U-2s was delivered; shortly the O3U-2 observation aircraft was changed to an SU-1 carrier scout designation.

VOUGHT O3U-3 (W.Larkins) The O3U-3 Corsair was a 1933 version still using the Pratt and Whitney Wasp engine but in a new version uprated from 450 to 550 horsepower. A floatplane gross weight of 4600 pounds still allowed battleship catapulting. A new tail shape distinguished this version and a ring cowl was added around the engine. Note the flotation bag container on the forward fuselage just outboard of the fuel tanks. Seventy six Wasp powered O3U-3s were built for the Navy.

VOUGHT O3U-3 (W.Larkins) Another view of a Wasp powered O3U-3 is shown, this aircraft from the battleship Colorado in the late 1930s, though at the moment without floats. The tail hook installation also ensured it was carrier-compatable. The new rounded tail with small dorsal fin shown shows up as does the single upper wing mounted machine gun and the flotation package. The O3U-3 had a high speed of over 160 mph.

VOUGHT O3U-3 (H.Thorell) An O3U-3 Corsair on floats attached to the cruiser USS San Francisco at a peaceful lake-front mooring, perhaps facilitating an officer's pleasant weekend. The aircraft appears to have been reworked to incorporate a special cockpit arrangement. As a floatplane the O3U-3, powered by the 550 horsepower Wasp, had a high speed of slightly over 150 mph and a gross weight of about 4600 pounds. Wingspan was the standard 36 feet of all the O2U- and O3U-Corsairs.

VOUGHT O3U-6 (Mfr.) After Vought devised an XO3U-5 with an experimental installation of a Pratt and Whitney Twin Wasp Junior engine in a new cowl, the next production observation Corsair was the O3U-6 of 1935. Shown at the factory before delivery to the US Marines on February 27, 1935 is one of 32 O3U-6s produced after the last O3U-3 was converted to an experimental XO3U-6. All these aircraft went to the USMC.

VOUGHT O3U-6 (Mfr.) Another photo (*above*) of the same O3U-6 Corsair on a snowy Connecticut airfield in February, 1935 shows it ready for delivery to Marine Squadron VO-8M without a plane number assigned. Still Pratt and Whitney Wasp powered with 550 horsepower available the 4750 pound O3U-6 had a high speed of 160 mph.

VOUGHT O3U-6 (Mfr.) The O3U-6 Corsair was tested as a seaplane though not so used by the USMC. The photo shows a sample of rough water testing required by the Navy on the same aircraft pictured earlier as a landplane. As a seaplane of 4900 pounds the O3U-6 dropped only a few miles per hour in speed capability.

VOUGHT XO4U-1 (Mfr.) A completely new light observation plane ordered from Vought in 1930, the XO4U-1 featured an all metal fuselage with the belly deepened to allow both metal framed lower and upper wings to tie in directly. This unusual design was not successful; the prototype crashed in 1931.

VOUGHT XO4U-1 (Mfr.) Another view of the unusual design aspects of the XO4U-1 prototype is given in a company photo. Note the inboard strut bracing from fuselage to lower wing. Powered by an uncowled Pratt and Whitney R-1340 Wasp engine of 500 horsepower, the XO4U-1 had a wingspan of 37 feet, a gross weight of about 3700 pounds, and a top speed of approximately 140 mph.

VOUGHT XO4U-2 (Mfr.) Though bearing no resemblance to the XO4U-1 except for the plane number, a new Vought observation type of 1932 was labeled an XO4U-2 even when it actually resembled Corsairs of the O3U- series. Featuring the Pratt and Whitney Twin Wasp Junior engine of 625 horsepower with individual exhaust stacks behind a neat long chord cowling, the plane was used only as a powerplant test bed. Gross weight was about 4700 pounds and high speed approximately 170 mph, but performance depended on the powerplant installed for test.

VOUGHT XO5U-1 (Mfr.) The Vought entrant in an early to mid-1930s competition for a new light shipboard-based observation type and pitted against the Curtiss XO3C-1 and Douglas XO2D-1, all powered by the 550 horsepower Pratt and Whitney Wasp engine. The original requirement was for an amphibian as shown in this Vought photo of May 8, 1934, the date of the XO5U-1 first flight. Note the original "park bench" type ailerons on the upper wing to allow more wing flap area, and also the cockpit canopy for the pilot.

VOUGHT XO5U-1 (Mfr.) Another view of the XO5U-1 prototype amphibian, this a Vought photo of May 14, 1936, two years after the aircraft first flight. The design of the amphibian landing gear is clearly shown and the large landing flaps on the upper wing are deflected. The earlier "park bench" ailerons have been changed to the more conventional inset type on both wings with a connecting drive strut used. All the experimentation came to little; the XO5U-1 was a losing competitor.

VOUGHT XO5U-1 (USN) The Navy withdrew the requirement for amphibious capability to allow the competing aircraft to come in at lower gross weight. The photo shows the XO5U-1 prototype using a main float without retracting wheels. A single .30 caliber gun is installed in the rear cockpit. The 550 horsepower Wasp engine powered the 4900 pound aircraft to a high speed of 155 mph. The competition was won by the Curtiss XO3C-1 and the XO5U-1 remained a single prototype and crashed in 1938.

CONVAIR OY-1 (F.Dean) The photo shows a rebuild of an OY-1 Sentinal in 1983, a World War II Marine Corps version of the Army Stinson L-5 liason airplane. A total of 306 OY-1 versions were obtained by the USMC and used in various forward area duties. Powered by a 185 horsepower Lycoming horizontally opposed engine, the OY-1 had a wingspan of 34 feet feet, a normal gross weight of just over a ton, and a high speed of about 125 mph.

PENNSYLVANIA AIRCRAFT SYNDICATE XOZ-1 (M.Sheppard) An experimental autogiro project with wings, the XOZ-1 was a modification of a Consolidated XN2Y-2 trainer into a rotorcraft on floats by the Burke Wilford Associates in Pa. It was notable as the first naval aircraft to fly with cyclic rotor control. Delivered to the Navy on August 9, 1957 at the Philadelphia Naval Aircraft Factory it was also the first rotary wing aircraft to fly off the water in the US. At forward speed lift could be unloaded off the rotor on to the wing.

NORTH AMERICAN ROCKWELL YOV-10A (P.Bowers) The North American entry in the Navy's Light Armed Reconnaissance Airplane (LARA) and declared the competition winner in August of 1964. Shown is one of seven YOV-10A Bronco prototypes, the first of which flew July 16, 1965. The development aircraft were powered by two 660 shaft horsepower Airesearch T76 turboprop engines. Wingspan was only 30 feet on the prototypes.

NORTH AMERICAN ROCKWELL OV-10A (Mfr.) Flight view of a production OV-10A Bronco shows the twin engines and boom and tail configuration. Production was ordered in October, 1966 and first flight took place in August, 1967. By late 1969 the USMC had obtained 114 Broncos after production ended in April. Wingspan had been upped to 40 feet and the T76 engines uprated to 715 horsepower each. With a crew of two, weapons could be carried on short sponsons off the fuselage as well as under it. Normal gross weight was 9900 pounds with overload capability to about 14500 pounds; high speed at sea level was about 280 mph without stores aboard.

NAVY OBSERVATION SCOUT AIRPLANES

In 1936-37 the Navy, in the multi-purpose philosophy of the time, set up a VOS airplane category combining the primary mission of observation with the secondary mission of longer range scouting. Typically the VOS class airplanes were first set up without the requirement for wing folding, and would go aboard battleships where there was presumably more space for catapult planes than on cruisers where VSO catapult planes with foldable wings would be used. Actually there were small differences in equipment and mission performance capabilities between the VOS and VSO airplanes.

The 1936 candidates for the first production VOS wheel-to-float convertable airplane were the Vought XOSU-1 Wasp powered biplane, a prototype rebuilt from a production O3U-6 Corsair with greater fuel capacity and full span flaps, a new all metal monoplane XOS2U-1 with a Wasp Junior engine, a Naval Aircraft Factory Wasp powered all metal biplane design labeled an XOSN-1,

and another biplane from Stearman, the XOSS-1, also using a Pratt and Whitney Wasp engine. The new Vought monoplane won the VOS competition and became a production item as the OS2U- Kingfisher series -1 to -3 for use on battleship catapults, and with either wheels or floats a Naval Inshore Patrol aircraft. The Kingfisher served throughout World War II as a major production naval type; the NAF supplemented Vought's OS2U- production with a similar OS2N-1 Kingfisher.

The final VOS airplane was an Edo Corporation design from the renowned float maker; the XOSE-1/-2 single or two place Ranger engine powered folding wing floatplane of all metal monoplane configuration. It flew shortly after the end of World War II when the requirement for the VOS type of airplane was to disappear. Other than construction of a few prototypes the XOSE- was not to make the grade as a service airplane. Innovations such as radar were to make the VOS airplane type obsolete.

EDO XOSE-1 (Mfr.) Dynamic just off the water photo of a single seat XOSE-1 observation scout seaplane by Edo, the premier float-maker, one of six prototypes with a first flight just after World War II in December of 1945. The XOSE-1 could also be used on wheels. A variant was the two seat XOSE-2. With war's end and the program going poorly the XOSE-1 was shortly cancelled. Powered by a Ranger V-770 engine of 550 horsepower, the seaplane had a normal gross weight of about 5300 pounds and a maximum speed of about 200 mph.

NAVAL AIRCRAFT FACTORY XOSN-1 (USN) One response to a 1936-37 Navy requirement for a new small observation scout plane convertable from floats to wheels was a single prototype from the Naval Aircraft Factory in Philadelphia. Ordered in 1936, the new two place biplane with metal fuselage, tail, and wing framing was characterized by a minimum of interplane struts, wing slats, and enclosed cockpits. The XOSN-1, shown in a Navy flight test photo of June 17, 1938, was powered by a 550 horsepower Pratt and Whitney Wasp engine, weighed 5500 pounds as a seaplane, and had a high speed of about 160 mph in either configuration.

NAVAL AIRCRAFT FACTORY XOS2N-1 (USN) During early wartime in World War II Vought was really stretched thin between delivering their OS2U- Kingfisher observation scouts and getting into production on their new F4U-1 Corsair fighter. The OS2N-1, shown in a March 2, 1944 Navy photo, was produced by the Naval Aircraft Factory in 1942. It was the equivalent of the OS2U-3 from Vought, and 300 aircraft were delivered from the Philadelphia factory to Naval Inshore Patrol Squadrons.

STEARMAN XOSS-1 (Mfr.) Another prototype intended as a ship-based observation scout came from Stearman as the XOSS-1 shown in landplane form in the photo. To lower landing speeds this Stearman biplane employed a full span flap on the upper wing with ailerons on the lower. Wingspan was 36 feet, gross weight as a seaplane about 5500 pounds, and high speed over 160 mph. Like the NAF XOSN-1 aircraft the Stearman used the 550 horsepower Pratt and Whitney Wasp engine, and also lost the competition.

VOUGHT XOSU-1 (Mfr.) Shown in a Vought photo of October 27, 1936 is an O3U-6 observation type revised to an XOSU-1 observation scout configuration. The unique arrangement of wing flaps and ailerons distinguished this model. Both upper and lower wing trailing edges had flaps and ailerons, the inner sections being flaps connected by a drive strut and outer ailerons also connected by another drive strut. It is not known whether the ailerons were also drooped as flaps in landing. Specifications were somewhat similar to those of an O3U-6, though the plane was heavier and slower. There was no production.

VOUGHT XOS2U-1 (Mfr.) The single prototype of a very successful observation scout type, this XOS2U-1 ship or shore based all metal monoplane is shown in a May 20, 1936 Vought photo, a day after its first flight; a landplane version had flown over two months earlier. With pioneering structural features and a strut and wire braced main float, the 4700 pound gross weight prototype was powered with a 450 horsepower Pratt and Whitney Wasp Junior engine, spanned 36 feet, and could attain a high speed of about 175 mph.

VOUGHT OS2U-1 (Mfr.) Successful testing of the prototype led to a Navy order of mid-1939 for 54 OS2U-1 Kingfishers which were delivered the next year. A landplane version is shown in this company photo of May 14, 1940, the first month of deliveries; one of two wing store stations is loaded with a 116 pound bomb. The production model added a third float-to fuselage support strut aft on the seaplane version. Fifteen of these aircraft went to NAS Pensacola, Florida and several went to Pearl Harbor. Others went aboard ship starting on the battleship Colorado in August, 1940. Landplane gross weight was about 4800 pounds; speed was approximately the same as that of the seaplane version.

VOUGHT OS2U-1 (USN) Flight photo of August 11, 1940 shows a production OS2U-1 Kingfisher landplane in service. Armament consisted of a single fixed .30 caliber gun forward and a rear flexible .30 caliber weapon for the rear seat man along with two wing store stations for light bombs or depth charges. A dozen OS2U-1s went to Alameda, Ca. in 1940. Spot welding was used extensively in fuselage construction. The wing had a single spar; the leading edge section was metal covered; fabric covering was used aft of the spar.

VOUGHT OS2U-2 (USN) The next Kingfisher version was the OS2U-2 with 158 ordered late in 1939, most of these delivered in late 1940 as landplanes but with float sets available. Those delegated to Inshore Patrol Squadrons would carry two depth charges, and the author well remembers them so loaded flying over his house from the Salem, Ma. USCG air base. The photo shows three of the 46 OS2U-2s that were delivered as seaplanes to NAS Pensacola, Fl.

VOUGHT OS2U-2 (USN) An OS2U-2 Kingfisher in a Navy photo show the seaplane configuration well. The -2 model was heavier, using protected fuel tanks and some armor. Over 50 OS2U-2s went to NAS Jacksonville, Fl. for Inshore Patrol duty. Kingfishers were becoming increasingly useful aircraft in many and varied missions.

VOUGHT OS2U-2 (F.Schertzer) A view of an OS2U-2 landplane in 1941 shows some details of a Kingfisher aileron and flap arrangement. Ailerons drooped to act as flaps on landing. The aircraft is plane number two assigned to the seaplane tender AV8, the USS Tangier. The "greenhouse" for the Kingfisher rear seat man was quite extensive. Hatches were slid forward to use the rear gun.

VOUGHT OS2U-3 (USN) Vought swung into high production with the OS2U-3 Kingfisher shown here on June 25, 1941 as a landplane just as production started. Over 1000 of the -3 version were turned out in a little over a year. A couple of OS2U-3s served with the Marines; most all went aboard Navy ships including battleships and cruisers of all types. The British received a total of 100 Kingfishers.

VOUGHT OS2U-3 (USN) An interesting view of a Kingfisher taxiing at NAS Jacksonville, Fl. in March of 1943. The -3 had additional wing fuel and increased armor protection. Its design was duplicated by the NAF-manufactured OS2N-1. A most publicized feat of many for a Kingfisher was the Pacific Ocean rescue of Captain Eddie Rickenbacker and crew downed at sea by taxiing across some 40 miles of rough water with men clinging to it. The 450 horsepower Wasp Junior engine gave the -3 a high speed of about 160 mph. Normal gross weight of the seaplane was three tons.

NAVY PATROL AIRPLANES

Early Navy patrol airplanes were flying boats with roots traced back to the early efforts of Curtiss and others on these water-based types. Practical flying boats started with Curtiss in the 1912-13 era before World War I. The first wartime uses of Navy boats were for anti-submarine warfare (ASW) against German U-boats. Flying boats dominated the patrol airplane (VP) area until World War II when a few landplanes crept in, and it was not long after World War II, perhaps a decade, that landplanes were used almost exclusively for Navy patrol missions because of their greater speed and range and endurance capability.

In the pre-World War I era a few Curtiss C boats were procured by the Navy. These single engine pusher biplanes were powered by Curtiss engines and pioneered Navy flying boat development.

In 1918 some single engine MF boats were used by the Navy as a larger more powerful development of the earlier Curtiss F boat, itself a development of the C boat.

Two other Curtiss Flying boat models were developed in World War I with their principal mission being ASW patrols off the European coastlines. The HS- boats were 1918 biplane types with a single Liberty engine driving a pusher propeller and having a crew of three. The HS-2L had increased wing area over the HS-1L to increase bomb load capability. The H- boats were twin engined developments of the pre-war Curtiss America boat intended for trying a war-interrupted trans-Atlantic flight. The H- series included the H-4 Small Americas and the later and bigger H-12s and H-16s as Large Americas. These patrol boats were used by both the British and the US in World War I and then into the 1920s.

The next improvement in patrol flying boats was the F-5L of 1918 built by both Curtiss and the NAF using two Liberty engines, basically a British Porte design and thus a combination of US and UK technology of the time. The F-5Ls were used into the late 1920s as a standard Navy patrol plane.

In late 1918 the famous big NC- "Nancy" three and four engined flying boats were built by Curtiss and the NAF. The NC- type was designed for trans-Atlantic capability to defeat the U- boat threat, and after the war in 1919 was the first type to accomplish that feat. These aircraft were used into the 1920s along with H-16s and F-5Ls.

In 1925 two new twin engined patrol boats appeared, the single Boeing PB-1 with pusher-tractor engines, first Packard water cooled models and later air cooled Hornet radials as a PB-2, and the seven place Naval Aircraft Factory PN-7 with twin tractor Wright liquid cooled engines. The PN-7 biplane was basically an improved F-5L boat and the first of a series of NAF-built patrol boat designs. The PN-8 added a metal hull, and the PN-9 made headlines in 1925 by both flying and floating "under sail" from the US west coast to Hawaii. More metal structure characterized the two PN-10 variants of the PN-9 and the PN-11 biplane boats of 1925 used two radial air cooled engines as did the last two PN-12s of 1931 which combined radial engines and a single vertical tail.

The production boat based on the NAF PN- series was the Douglas PD-1 patrol plane with 25 aircraft delivered in 1929. It used two tractor mounted Wright Cyclone radial engines. Sikorsky started making Navy patrol boats based on their civil S-38 sesquiplane amphibian in 1927 as a few PS- types with twin engines and gunner's positions; these planes were later modified to RS- transport aircraft with armament removed.

The year 1929 saw a major flying boat design development in the Consolidated 100 foot wingspan parasol wing XPY-1 Admiral monoplane with two Wasp radial engines and a crew of five in open cockpits. It was the fore-runner of a long line of Consolidated and Martin patrol boats. It was tested with three engines as well, but that arrangement was not a success.

Hall Aluminum started their line of twin radial engined biplane PH- patrol boats in 1930 with the last of these delivered in 1939 and serving into early World War II. Also in 1930 Martin made production PM-1 and PM-2 boats based on the NAF PN-12 design. Another twin Cyclone radial engined biplane boat, there were a total of 55 manufactured with the -2 model having enclosed cockpits. The NAF put out the single XP4N-1 twin Cyclone powered patrol boat prototype in 1930, one of the last of the NAF-designed boats; it was followed by two -2 prototypes leading to the Hall boat design.

In 1931 Keystone delivered 18 twin engined open cockpit biplane boats with twin vertical tails to the Navy as PK-1s, conventional for the period and based on NAF designs. The same year Martin put out their XP2M-1 similar to the XPY-1 of Consolidated with twin engines and also tested with a third. Soon after Martin garnered the production contract for a 100 foot span flying boat over Consolidated with their P3M- boats which lasted into 1937 in Navy squadrons. Also in 1931-32 the Douglas P2D-1 appeared, another twin engined biplane originally intended as a T2D-1 torpedo landplane or as a bomber. Army disapproval of this Navy version resulted in the type going over to a twin float equipped patrol seaplane operating in VP squadrons.

The year 1932 saw Hall Aluminum produce a single large four engine prototype twin tailed biplane flying boat, the XP2H-1. The type was unusual in using liquid cooled Curtiss Conqueror engines. In addition Fokker produced a few twin pusher engined PJ-1 monoplanes as "Flying Lifeboats" for the US Coast Guard. The prototype XP2S-1 from Sikorsky was also delivered for test in 1932 as a twin engined biplane with a unique pusher-tractor powerplant arrangement and two gunner positions. More importantly Consolidated unveiled their new XP2Y-1 Ranger prototype leading to P2Y-1 to -3 twin Cyclone powered models in service from 1933 through the remainder of the 1930s both in frontline and later in training service.

A 1935 twin engined metal monoplane patrol flying boat competition between the XP3D-1 prototype from Douglas and the XP3Y-1 from Consolidated resulted in a win for the XP3Y-1 which was then developed into the famous PBY- patrol bomber series.

The first of the landplane patrol planes was the 1942 twin engined Lockheed/Vega PV-1 Ventura, a Navy version of Lockheed's civil Model 18 Super Electra used extensively in World War II and followed by a further developed PV-2 Harpoon variant. Another 1942 patrol plane prototype was the new Consolidated Model 31 civil flying boat converted and revised to the single navalized armed XP4Y-1 Corregidor, but there was no production.

The next aircraft in the Navy patrol plane category were both 1945 landplanes, the (second) Boeing PB-1, simply a variant of the Army B-17 Flying Fortress bomber with armament stripped out and radar and sometimes a lifeboat added, and the first of the early warning types as a PB-1W, and the Lockheed twin engined P2V-Neptune, a highly efficient specially designed from the start aircraft for naval patrol duties and just missing World War II. Production Neptunes in many forms, including those with jet engine assistance for dash speeds served the post-war Navy in active front line squadrons and then in the Reserves right into the 1970s.

Late 1940s Navy patrol planes included two more landplanes, the 1947 limited production Martin P4M-1 Mercator equipped with two piston engines and two jets for dash speed boost, and the 1949 Lockheed civil 749 Constellation converted with warning radar equipment as a Navy PO-1W, later redesignated as a WV-1 early warning airplane.

The 1950s saw, in 1951, the remaining Navy wartime Convair PB4Y-2 Privateer four engine landplane patrol bombers redesignated as P4Y-2 patrol planes since there was no bomber mission for them, and in the same time period two new patrol seaplanes were in development, the Martin P5M- Marlin and the Convair XP5Y-1. The twin piston engined Marlin was developed from the PBM-5 Mariner and was further modified into a P5M-2 with extensive ASW gear. The Marlin was the last production Navy flying boat. The XP5Y-1 was an ambitious four engine turboprop flying boat, heavily armed, which was later redesigned into the R3Y-1 transport boat, but powerplant problems made development lag and there was no production.

In 1956 a final try at an advanced patrol flying boat was made by the Navy and Martin with the XP6M-1 jet powered craft, the four engined Seamaster with a swept wing and advanced hull design. Of the several prototypes two crashed, and by 1956 the program was ended.

The final patrol plane in the long series was the current four turboprop engined Lockheed Orion, a 1961 navalized version of the civil Electra transport designated originally as a P3V-1 and later a P-3 in A, B, and C versions. The P-3 is the Navy's land based ASW aircraft. New ASW patrol types have been proposed, but no new prototypes have been built and tested to date.

CURTISS C-2 (Mfr.) Shown hauled up on shore at Lake Keuka in Hammondsport, NY. in 1913 the Curtiss Navy C-2 flying boat was one of five more or less similar aircraft purchased by the Navy. These boats were based on the civil Curtiss F boat design. The C-2 was used during tests of a Sperry autopilot. In 1914 the C-2 was redesignated as an AB-2. Wingspan was about 42 feet. A 75 horsepower Curtiss Model O engine powered the C-2 to a high speed of about 55 mph. Note the "USN" on the hull.

CURTISS C-5 (USN) Another of the early Navy C boats, this apparently the C-5 and the last of the five boats purchased. Note the anchor on the rudder. This plane is an equi-span naval version of the Curtiss civil F boat. Noteable are the mid-gap ailerons, typically early Curtiss, and the sloped Goodier strut running from the forward hull to the engine mount to protect the pilot from the engine in a crash.

CURTISS MF (Mfr.) A 1918 successor to the early Curtiss F boat, the MF evolved after several experimental flying boats had not been successful. A two place single engine pusher like the earlier F, the MF had a modernized hull with sponsons and an uprated Curtiss OXX engine developing 100 horsepower thus providing a high speed capability of over 70 mph. Wingspan was about 50 feet and gross weight 2500 pounds. Curtiss built over 20 for the Navy; the Naval Aircraft Factory built about 80.

CURTISS HS-1L (USN) A larger single engine three place biplane boat of 1917-18 in the Curtiss mold using a pusher version of the 360 horsepower Liberty engine, this aircraft had a nose gunner up front with a Lewis gun or two and the under-wing capacity for 360 pounds of depth charges for anti-submarine warfare. Several hundred were built both by Curtiss and Lowe Willard Fowler. Span was 62 feet, gross weight about three tons, and a high speed of about 85 mph could be achieved. Navy HS-1L aircraft were operated from naval stations on the French coast in World War I.

CURTISS HS-2L (Mfr.) In order to increase bomb carrying capability Curtiss modified the HS-1L design into an HS-2L by increasing the span of both wings by 12 feet. Changes were also made in the vertical tail. The Curtiss photo of November 16, 1918 shows an HS-2L with a bomb under the port wing. HS-2Ls were used for ASW patrols starting in the spring of 1918. In a coordinated effort of about a half dozen manufacturers several hundred HS-2Ls were turned out. Winspan was 74 feet, gross weight increased several hundred pounds over the HS-1L, and the same Liberty engine gave a high speed of slightly over 80 mph.

CURTISS HS-2L (USN) A Curtiss-built HS-2L sits on its dolly showing the forward hull sponsons and the forward cockpit for the nose gunner. Besides Curtiss HS-2Ls were built by Boeing, Lowe Willard Fowler, and Gallaudet and others. These flying boats stayed in service for the Navy well into the 1920s, and some were assembled from spare parts at the Naval Aircraft Factory.

CURTISS H-12 (Mfr.) The first twin engined four place Curtiss flying boats were in the H- series. The H-4 Small Americas were built for Britain. A later and larger version was the H-12, shown in the photo of the prototype aircraft just after launch. Deliveries of 20 were made to the Navy in 1920 after a large number of similar Large Americas had gone to Britain. Early H-12s, including this prototype, were powered by Curtiss V-2 engines of 200 horsepower each. Wingspan was about 92 feet, gross weight over five tons, and high speed about 80 mph. Armed with up to four flexible Lewis guns in two positions, the H-12s were later refitted with 360 horsepower Liberty engines and then redesignated H-12Ls.

CURTISS H-16 (Mfr.) A later and final version of the Curtiss H-series of twin engined flying patrol boats, the H-16 was just about the size and weight of the H-12, but eventually used a more powerful 400 horsepower version of the Liberty engine, and was faster at over 90 mph top speed. The Curtiss photo shows an H-16 ready to receive its tractor Liberty engines. The enclosed pilots' cockpit of this four place boat is noteable.

CURTISS H-16 (USN) A Navy photo gives a good April, 1925 flight view of a special H-16 boat fitted out for use by Admiral Moffat. Curtiss built over 180 H-16s; the NAF built many others including the one pictured. The first H-16 came out in mid-1918 and lasted well into the 1920s.

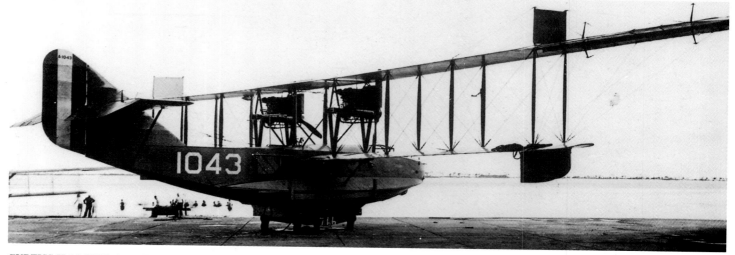

CURTISS H-16 (USN) A good view of a Curtiss-built H-16 on its handling dolly on August 23, 1918, only two months after the first H-16 aircraft appeared. Wingspan was 95 feet, gross weight about five and one half tons. Normally the four place aircraft had a gunner at the nose position, two pilots side by side well forward of the propellers, and an aft upper gunner's position. Note the side window in the hull.

CURTISS F-5L (USN) A design of mixed ancestry, being derived from a British Porte-designed hull combined with other Curtiss H-series type elements, an F-5L (the L for Liberty engines) is pictured as a command transport on July 7, 1927. Instead of the enclosed pilot's cabin of the H-16 the F-5L had open cockpits. Both Curtiss and the Naval Aircraft Factory built F-5L boats; first F-5L flight took place in July, 1918 at the NAF which built 137. Curtiss built 60 more. Like the H-16 boats many lasted well into the 1920s. Wingspan was 104 feet, gross weight normally about 13500 pounds, and like the H-16 with Liberty 12A engines of 400 horsepower top speed was about 90 mph.

CURTISS NC-4 (USN) Flight photo of the most famous of the NC-"Nancy" big flying boats, this fame gained by being the first aircraft to fly across the Atlantic Ocean in 1919. These boats originated in response to the German submarine threat in World War I. The ability to fly the Atlantic meant such a plane could effectively counter such a menace anywhere. The NC- designation meant Navy-Curtiss, since the big boats were the result of very close cooperation between that service and the experienced flying boat contractor.

CURTISS NC-1 (Mfr.) The NC-1 "Nancy" is shown on its handling gear in original configuration with one of its three tractor Liberty engines running up its four blade propeller. A gunner location is at the extreme nose of the hull; pilots were located at the rear of the center engine nacelle high up between the wings. NC-1 flew first in October, 1918, just before the end of World War I. Later the NC-1 was modified with NC-2 wings and a fourth engine as a centerline pusher for the May, 1919 trans-Atlantic attempt; it went down at sea and sank. The crew was rescued.

CURTISS NC-3 (Mfr.) This photo of the NC-3 of late April, 1919 shows it in original configuration with four Liberty engines in a two-nacelle tractor-pusher arrangement driving two bladed propellers. Note the short hull and braced tail design of the NC-boats. Wing span was 126 feet; gross weight about 14 tons, and the four Liberty 12 cylinder engines could drive the flying boats to a high speed of over 80 mph. Like the other boats the NC-3 was modified in engine arrangement for the trans-Atlantic attempt. It was forced down at sea like the NC-1, but managed to taxi safely to the Azores.

CURTISS NC-4 (Mfr.) Another view of the aircraft that made it across the Atlantic, the NC-4, shows the T/A engine arrangement of a central twin powerplant nacelle with both a tractor and pusher installation along with the two outboard tractor engine/propeller nacelles. A rear gunner's position was aft on the hull. The short hull design of the Navy turned out to be successful. The NC-4 aircraft is still preserved at the Naval Aviation Museum.

NAVAL AIRCRAFT FACTORY NC-6 (USN) Additional NC- boats were later produced at the NAF in Philadelphia, Pa. with a half dozen fabricated. The bracing for the high tail is well shown here. The NC-6, along with NC-5, had three engines, the outboards being tractor and the center one a pusher powerplant. A normal crew was five.

NAVAL AIRCRAFT FACTORY NC-9 (USN) A Navy photo at the NAF shows an excellent view of the NC-9 on April 11, 1922. Behind it sits an NAF -built TS-1 fighter on floats. The NC-9 had the same four engine arrangement as the NC-4. By the mid-1920s the NC- boats were gone.

NAVAL AIRCRAFT FACTORY NC-10 (USN) The final "Nancy" boat, NC-10 at its mooring on July 25, 1921 shows it had the same four engine arrangement as the NC-4 with the center nacelle having both tractor and pusher engine installations. The maze of struts and wires on the big flying boat is readily apparent.

BOEING PB-1 (Mfr.) The single PB-1 was built in 1925 against a Navy requirement for a flying boat capable of making it from the US west coast to Hawaii non-stop. Construction was mostly metal but the top of the hull was wood and wings and tail were fabric covered. Two 800 horsepower Packard 2A-1500 engines in a tandem tractor-pusher arrangement provided the power, and gave a high speed of 125 mph. Span was 87 feet six inches and gross weight 23500 pounds. The PB-1 could carry 2000 pounds of bombs (though not to Hawaii). There were problems with the Packard engines and the plane later was modified as a PB-2 fitted out with Pratt and Whitney Hornet radial air cooled engines.

BOEING PB-1G (Mfr.) Neglecting the fact that the designation had previously been used by another earlier Boeing plane, the US Coast Guard version of the B-17G Flying Fortress was called a PB-1G, the G for Coast Guard. Starting in 1945 over a dozen B-17s were converted with armament removed and radomes in place of the forward chin turret. The USCG used its Fortresses for air-sea rescue where a lifeboat could be carried under the fuselage. The power of four 1200 horsepower Wright R-1820 Cyclones gave the 55000 pound PB-1G a maximum speed of over 300 mph. Some other Fortresses were converted to PB-1W airborne early warning versions with a belly radar.

DOUGLAS PD-1 (H. Thorell) Douglas got into the naval patrol flying boat area in 1927-28 when it produced 25 twin tractor engined biplanes. Forward and aft flexible gun positions were located in the hull and about 1000 pounds of bombs could be carried underwing. The photo shows a PD-1 in Hawaii on July 10, 1933, four years after entering service. Using Wright radial engines each of 525 horsepower the 15000 pound gross weight PD-1s had a high speed of about 120 mph.

DOUGLAS P2D-1 (Mfr.) A company photo of a Douglas P2D-1 landplane before delivery to the Navy shows a large twin radial engined biplane. Capable of carrying a torpedo or 1600 pounds of bombs from land, it appeared to be a Navy intrusion on an Army mission. To keep inter-service peace the aircraft, 18 of which were ordered in mid-1930 and delivered in 1932, were equipped with twin floats and given a Navy patrol designation. They operated around the Panama Canal. Powered by Wright R-1820 Cyclones, the landplane version had a high speed of almost 140 mph.

DOUGLAS P2D-1 (USN) A P2D-1 of Navy Patrol Squadron VP-3 is shown on twin floats on May 10, 1933, probably in the Canal Zone area. These big floatplanes with a wingspan of 57 feet and a gross weight of about 12700 pounds had a high speed of 130 mph. Service ceiling was about 10000 feet. Armament consisted of two flexible .30 caliber guns , one in a forward cockpit and one aft. These aircraft served until early 1937.

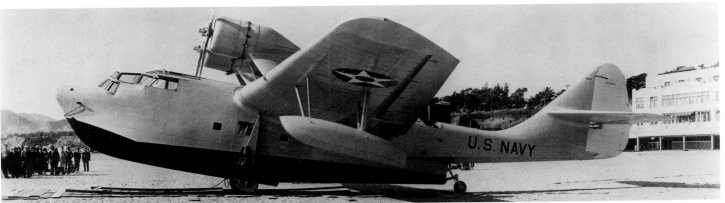

DOUGLAS XP3D-1 (Mfr.) The first monoplane flying boat design by Douglas, the twin engined XP3D-1 prototype appeared early in 1935 as a competitor to the Consolidated XP3Y-1 to replace the Navy's fleet of biplane patrol boats. Two 825 horsepower each Pratt and Whitney Twin Wasp engines were pylon mounted well above the high monoplane wing. A closed cabin was provided for the pilots and major structure was of metal. Wingspan was 95 feet, gross weight about 22000 pounds, and high speed approximately 160 mph.

DOUGLAS XP3D-2 (Mfr.) After early testing and with an eye on the competition Douglas revised its design considerably in 1936. The XP3D-1 prototype was updated to an XP3D-2 model with Twin Wasp power increased to 900 horsepower each, and engines were lowered to a wing position giving less drag, and tip floats retracted as on the XP3Y-1 competition. Gross weight rose to about 22800 pounds and high speed rose to about 20 mph greater. The XP3Y-2 lost the competition, however.

HALL XPH-1 (USN) Shown armed with twin Lewis machine guns fore and aft in two gunner's cockpits, the Hall Aluminum XPH-1 biplane flying boat of late 1929 was an all metal hulled improvement of previous Naval Aircraft Factory designs. Powered with two carefully cowled Wright R-1570 radial engines of 500 horsepower each, the new Hall boat had a high speed of about 125 mph.

HALL XPH-1 (Wright) Drawn up on the ramp with beaching gear in place, the XPH-1 gives an excellent view of the nose gunners ring mount, less gun, the side-by-side open pilots' cockpits with headrests, and details of powerplant installations along with three bladed Hamilton propellers. After tests of this prototype nine PH-1s with revised ring cowlings over Wright R-1820 engines were ordered in 1930. The PH-1s served until 1937 and were somewhat heavier but about 10 mph faster than the prototype.

HALL PH-2 (P.Bowers) One of seven improved Hall twin engined flying boats, the PH-2s were ordered for the US Coast Guard in 1936. Little different from the PH-1s except for the two Wright Cyclones being uprated to 750 horsepower each and enclosed pilots' cockpits, the PH-2 was somewhat faster because of the installed power increase. These Halls served right up until the US entered World War II.

HALL PH-3 (P.Bowers) The final Hall boat (*above, right*), the last "production" biplane patrol plane, was the PH-3. Seven of these Halls were procured in 1939, again for Coast Guard air-sea rescue and patrol work. Changes apparent were new engine cowlings for the Cyclones and a slicked up pilots' enclosure. Again slightly heavier than earlier models, the PH-3s lasted long enough to serve in early wartime.

HALL XP2H-1 (USN) Ordered as a prototype by the Navy only a few days after they gave Hall a production order for twin engined PH-1s, the large XP2H-1 four engine biplane flying boat was somewhat an anachronism for 1932 when first flown. Shown in a Navy photo of November 16, 1932, a day after its initial flight, the XP2H-1 was powered by liquid cooled 600 horsepower Curtiss Conqueror engines configured using two large nacelles each with a pusher and a tractor propeller installation. As seen in the photo a large part of the nacelle consisted of cooling radiators. With five gun positions, a 112 foot wingspan, and a gross weight of **35000** pounds, the XP2H-1 was a sizeable aircraft with a high speed of about 140 mph. It remained a prototype however.

FOKKER PJ-1 (H.Thorell) Designed by American Fokker and produced only after Fokker was absorbed by General Aviation, the five four place monoplane PJ-1 "Flying Lifeboats" went to the US Coast Guard in 1932, in fact for a time they were often called FLBs. These boats served faithfully until the late 1930s and the author remembers seeing one at the Salem, Ma. USCG Station with the unusual high pusher propeller design. Powered by two Pratt and W hitney Wasp engines of 420 horsepower each, the 11200 pound gross weight PJ-1 had a high speed of about 125 mph. One of the five was changed to a tractor propeller arrangement and labeled a PJ-2.

KEYSTONE PK-1 (USN) The only Navy patrol flying boat design built by Keystone-Loening of Bristol, Pa., the PK-1 twin engine biplane boat was derived from NAF designs. A total of 18 PK-1s were ordered in 1929 and delivered in 1931. The Navy photo shows a PK-1 of Patrol Squadron VP-1 based in Hawaii on February 24, 1933. Two nicely cowled Wright Cyclone engines each of 575 horsepower driving three bladed propellers powered the 16500 pound gross weight PK-1 to a high speed of 120 mph. Wingspan was 72 feet. Well braced twin vertical tails were a characteristic of this Keystone flying boat.

MARTIN PM-1 (P.Bowers) A moored example of a Martin PM-1 patrol boat from Squadron VP-9 is shown. The PM-1 aircraft were very similar to both the NAF PN-12 and Douglas PD-1 twin engined flying boats with the PN-12s being prototypes. Distinguished by a broad single vertical tail, the example shown is a boat modified with ring cowlings over uprated Cyclone engines and enclosed pilots' cockpits. A total of 30 PM-1s were built with a 1929 order being filled the next year.

MARTIN PM-1 (P.Bowers) Another example of a PM-1, this from Squadron VP-7, rides at a mooring. The original PM-1s used two 525 horsepower uncowled Wright radial engines but were later updated to use 575 horsepower ring cowled Wright Cyclones where gross weight increased from about 15400 to nearer 16100 pounds. Wingspan was 72 feet ten inches, and maximum speed between 115 and 120 mph. PM-1s were used in Navy Patrol Squadrons VP-7, -8, and -9.

MARTIN PM-2 (USN) A batch of 25 modified Martin PM- boats were ordered in 1930 as PM-2s with the more powerful cowled 575 horsepower Wright R-1820 Cyclones and enclosed cockpits but using a twin vertical tail arrangement with plenty of bracing struts like the Keystone PK-1 and some of the Naval Aircraft Factory boats. The PM-2 pictured on April 3, 1936 was one of those used by Utility Squadron 1. PM-2 specifications were quite similar to those of the PM-1 except gross weight was up to about 17000 pounds. A few PM-2s lasted at least until 1937.

MARTIN XP2M-1 (USN) Built under a 1929 Navy prototype development contract issued 16 months after that for the Consolidated XPY-1 Commodore type, both for a 100 foot wingspan flying boat, the Martin XP2M-1 flew in 1931. Basically a Consolidated design, there were differences, particularly in engine type and placement. Shown in a Navy photo of June 6, 1932 as a twin engined aircraft with ring cowled 575 horsepower Wright R-1820 Cyclones nicely lined up with the wing in an optimum low drag position, the XP2M-1 was initially tested with a third Cyclone centrally strut-mounted well above the wing. In this configuration gross weight was about 10 tons and high speed about 140 mph.

MARTIN XP2M-1 (Mfr.) The XP2M-1 prototype is shown in flight with the third engine high above the wing in a 1931 test. Obviously there was a gain in speed and a loss in range capability. Eventually the Navy selected the longer range twin engine version. The arrangement of outboard stabilizing floats was new and enclosed cockpits were now the rule for the new monoplanes. Tail design needed some work, however!

MARTIN P3M-1'(USN) After Martin won the production contract for the 100 foot wingspan monoplane flying boat over Consolidated in June, 1929, nine 1931-32 Martin production versions of the Consolidated XPY-1 boat were designated P3M-1. Strangely, perhaps, Martin kept the un-cowled engines strut-supported below the wing like the XPY-1, but the boat had only two engines. Powered by two Pratt and Whitney Wasps of 450 horsepower each, the approximately eight ton gross weight boats had a top speed of about 115 mph.

MARTIN P3M-2 (USN) The P3M-1s were later converted to a P3M-2 configuration as shown in this Navy photo of September 14, 1933. Major change was conversion to 525 horsepower R-1690 Hornets using ring cowls resulting in a slight high speed increase. The nine aircraft were still with VP-15 in mid-1937, but were replaced by P2Y-2s in 1938.

MARTIN P4M-1 (P.Bowers) Martin entered the landplane patrol aircraft field in wartime 1944 with a Navy contract for two XP4M-1 prototypes distinguished by a mix of both piston and jet engines in two powerplant nacelles, the two Pratt and Whitney R-4360 piston engines used for cruising with the jets fired up in addition for dash speeds. The photo shows the armament of two 20mm cannon in each of nose and tail turrets and .50 caliber machine guns in a dorsal turret position. In addition a .50 caliber gun was located on each side of the fuselage. Two torpedoes could be carried for about 3000 miles.

MARTIN P4M-1 (R. Besecker) Clean lines of the P4M-1 Mercator show in this photo as well as details of nacelle and landing gear. First flight of a prototype XP4M-1 took place in September, 1946. It took awhile, but finally in the mid-1950s 19 production aircraft started coming from the Martin plant near Baltimore, Md. The 4600 pound thrust each Allison J33 jet engines were mounted aft in the nacelles. Inlet doors could be closed off for low drag in cruise flight; jet exhaust exited at the aft end of the nacelle when those engines were used. All P4M-1 Mercators went to Navy Squadron VP-21 and were used for several years.

MARTIN P4M-1Q (R.Besecker) Some of the Mercators were modified into P4M-1Q electronic countermeasures aircraft with specialized mission antennas added on the outside. Shown in the photo is a P4M-1Q operated by Navy Squadron VQ-1 in Japan. The Mercator had a wingspan of 114 feet. With only the two Pratt and Whitney R-4360 piston engines operating the aircraft had a top speed of just under 300 mph; with the jet thrust added high speed was about 415 mph. Normal gross weight was about 41 tons.

MARTIN XP5M-1 (Mfr.) The prototype for a new postwar Navy flying boat, the XP5M-1 from Martin is shown in a company photo. The new boat utilized a PBM-5 Mariner wing combined with a new hull design having improved aerodynamic and hydrodynamic qualities. The photo shows 20mm cannon turrets in nose, tail, and a dorsal position along with a radome above the enclosed cockpit area. Ordered in mid-1946, the XP5M-1 flew just under two years later. Powered by two Wright R-3350 Duplex Cyclone engines each of 2700 horsepower, the new patrol boat had a 30 ton gross weight and a high speed of about 250 mph.

MARTIN P5M-1 (Mfr.) A flight photo shows the clean lines of a production P5M-1 Marlin flying boat, of which a total of 114 were ordered from 1951 to 1954. Several design changes were made in the production type including nose turret replacement with radar and elimination of the dorsal turret, new wing floats and support struts, and ASW stores space provided in aft nacelle "bomb bays". Some P5M-1s served during the Korean conflict. Using essentially the same wing, the type had uprated Wright engines of 3250 horsepower each giving the 36 ton gross weight P5M-1 a top speed of about 260 mph. In 1962 P5M-1s became P-5As.

MARTIN P5M-1G (P.Bowers) Seven Marlin boats were provided the US Coast Guard as P5M-1Gs like that shown in the photo. A large boat, the Martin had a wingspan of 118 feet and a hull length of 91 feet. Note the dual wheel main and tail beaching gear. Handling a large flying boat was no easy task. The P5M-1G carried no armament, however, one of the aft nacelle stores compartments can be seen in the photo. The aft turret used in the Navy version was converted into an observer's station.

MARTIN P5M-2/P-5B (Mfr.) A Martin flight photo shows the final Marlin flying boat version, the P5M-2 appearing in 1954 with the most apparent change a new T-tail assuring the horizontal tail surfaces would stay clear of both water and wing wake. A total of well over 100 P5M-2 Marlins were delivered in the next six years; these were re-named P-5Bs in the 1962 redesignation scheme. These aircraft were designed primarily for the ASW mission, but a few were used in the Vietnam conflict.

MARTIN P5M-2S/SP-5B (R.Besecker) By the time this photo was taken of a Marlin of Squadron VP-40 in August of 1968 it was designated an SP-5B for anti-submarine use. The doors of the port nacelle weapons bay for torpedoes or other stores are open. Additional wing store stations appear just outboard and a searchlight is at the wingtip. The SP-5B was the final Navy production flying boat; landplanes were to take over after 1967.

MARTIN P5M-2S/SP-5B (R.Besecker) Another photo of the SP-5B Marlin of Squadron VP-40 taken after the big boats were pulled from service. A radome is over the pilot's compartment and magnetic anomoly detection (MAD) gear is located at the T-tail intersection. The Wright R-3350 engines each developed a maximum of 3700 horsepower giving the 39 ton gross weight Marlin a high speed of about 260 mph. Normal cruise speed was about 100 mph slower.

MARTIN XP6M-1 (Mfr.) The final flying boat developed for Navy use, the five place Martin XP6M-1 swept wing jet powered Seamaster design of 1958-59 was an attempt to get landplane performance in combination with flying boat basing versatility. Shown in a company photo coming up the ramp on beaching gear, the four engine XP6M-1 was loaded with new features including a water-tight weapons bay in the hull. Powered by Allison J71 jet engines with afterburning thrust of 13000 pounds each, the XP6M-1 had a span of 103 feet, length of 134 feet, weighed up to 95 tons fully loaded, and had a high speed of about 650 mph.

MARTIN XP6M-1 (USN) The Seamaster in flight was certainly a sleek design for a flying boat as seen in the Navy photo of September 11, 1956, 14 months after first Seamaster flight. Two XP6M-1s and several P6M-2s with uprated engines were fabricated, but after two crashes the program was cancelled. The US Navy was out of the new flying boat business in 1959 and Martin built no more airplanes.

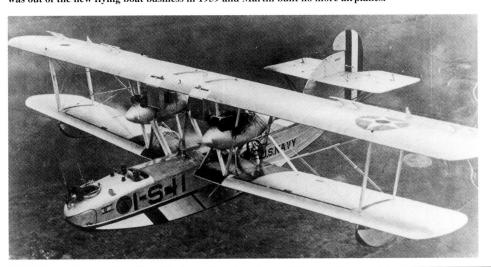

NAVAL AIRCRAFT FACTORY PN-7 (USN) The five place PN-7 shown in an April, 1925 photo was the first of a series of Naval Aircraft Factory biplane flying boats in the patrol category, and was called a PN-7 because it was a redesignated F-7L, itself an improved F-5L boat. The two PN-7s had a modified F-5L wooden hull with a new thick airfoil wing of 72 feet 10 inches span. Powered by two between-wings 525 horsepower each Wright T-2 liquid cooled engines with radiators above, the PN-7 had two Lewis guns and a 1000 pound bomb load could be carried. Gross weight was 14200 pounds and high speed 105 mph. A single PN-8 boat was similar except for a metal hull in place of wood.

NAVAL AIRCRAFT FACTORY PN-9 (USN) A Navy photo of July 29, 1925 shows the single PN-9 designed with sufficient fuel capacity to fly from the US west coast to Hawaii non-stop. The 1925 attempt to do so ended in the PN-9 coming down in the water many miles short after 28 hours in flight, but floating the rest of the way with make-shift sails. Powered by two 475 horsepower Packard 1A-2500 liquid cooled engines, the 19600 pound gross weight PN-9 could fly at a maximum speed of over 110 mph. Wingspan was the same 72 feet ten inches of the previous PN- boats.

NAVAL AIRCRAFT FACTORY PN-10 (USN) A Navy photo of July 30,1926 shows one of two PN-10 patrol boats made at the Naval Aircraft Factory in Philadelphia. Similar to the PN-9 except with less range capability and a metal instead of a wooden structure, and powered by two 500 horsepower Packard 2A-1500 engines driving three blade propellers, the PN-10s had the same 72 foot ten inch wing span and 49 foot length as earlier models, a gross weight of 19000 pounds and a high speed of 114 mph.

NAVAL AIRCRAFT FACTORY PN-11 (USN) A May 27, 1929 Navy photo shows one of four NAF PN-11 five place biplane patrol boats sporting a new simplified hull design and using radial air cooled engines. In addition twin heavily strut braced vertical tails sat in the slipstream of the two propellers. The initial PN-11 used two Pratt and Whitney 525 horsepower Hornet engines; the other three used Wright R-1750 engines of 500 horsepower each. The 17000 pound PN-11 had a high speed of over 125 mph.

NAVAL AIRCRAFT FACTORY PN-12 (USN) Originally ordered as PN-10s, the two PN-12s, one of which is shown in a Navy photo of April 24, 1931, were the same size as previous PN- patrol boats, and utilized two 525 horsepower Wright R-1750D radial engines driving three blade tractor propellers. One PN-12 was used as boat number 10 of Squadron VJ-1 as shown. Gross weight was 14100 pounds and high speed 114 mph. The PN-12 was the final design of the PN- series and had a single vertical tail.

NAVAL AIRCRAFT FACTORY XP4N-1 (USN) The one and only XP4N-1 of 1930 from the NAF was pretty much the same as the PN-12 except for reduction gearing in the Wright R-1750 525 horsepower air cooled radial engines. The plating of the all metal hull shows up here. The wide sponsons of the earlier boat hulls have been eliminated. The XP4N- boats retained the 72 foot 10 inch wingspan of the earlier PN-s. Gross weight of the XP4N-1 was just under 17000 pounds and high speed was almost 120 mph.

NAVAL AIRCRAFT FACTORY XP4N-2 (USN) One of two NAF XP4N-2 prototype biplane patrol boats, and the last NAF design leading to development of the Hall XPH-1 and production PH-1s. Again powered by two Wright radial engines of 525 horsepower each, but with more fuel capacity, the 17100 pound gross weight plane had about the same performance as the -1.

LOCKHEED PO-1W (Mfr.) A Lockheed photo shows one of two initial conversions of the civil 749 Constellation to a Navy patrol aircraft. First flown in this configuration with big radomes installed both above and below the fuselage, it was later redesignated as a WV-1 (the P for patrol changed to W for early warning missions, and the O changed to V to re-designate Lockheed.) The WV-1 was powered with four Wright R-3350 Duplex Cyclone engines each of 2500 horsepower maximum providing a high speed of about 330 mph. Wingspan was 123 feet and normal gross weight about 50 tons.

SIKORSKY XPS-1 (H.Thorell) The initial navalized version of the civil Sikorsky S-38 procured in 1927-28 as a patrol plane with a nose gunner position. Only one prototype was obtained powered with two Wright J-5 Whirlwind engines each of 220 horsepower giving the 8200 pound gross weight twin boom and tail sesquiplane amphibian a high speed of about 110 mph. The aircraft was later revised as an RS-1 transport type.

SIKORSKY PS-2 (H.Thorell) The Navy ordered two additional versions of the ten place S-38A in late 1928 and labeled them as PS-2s, but first as XPS-2s as shown on the tail of the plane in the photo. These cabin amphibians were later used by utility squadrons. Major difference from the PS-1 was use of two Pratt and Whitney R-1340 Wasps each of 450 horsepower giving the five ton gross navalized S-38A a high speed of about 125 mph. Wingspan was 71 feet eight inches. THey later became RS-2 transports.

SIKORSKY PS-3 (H.Thorell) This view of one of four Navy Sikorsky PS-3s of 1929 shows the two gun positions , one in the nose and one in the tail end of the hull. Similar to the earlier PS-2s except for revised Pratt and Whitney Wasp engines, the -3 models corresponded to the civil S-38B. Weights and performance were generally similar to the PS-2s. They later became RS-3 transports.

SIKORSKY XP2S-1 (USN) A final Sikorsky patrol boat prototype ordered in 1930 and delivered in mid-1932 was the single four place XP2S-1 biplane. Shown in a June 14, 1932 Navy photo, the XP2S-1 was unique in being a twin engine tractor and pusher 450 horsepower Wasp powered boat. A nose and aft gunner and a bomb underwing show the boat to be warlike. Gross weight was about 9500 pounds and wingspan 56 feet. The XP2S-1 had a high speed of about 125 mph. There was no further procurement.

LOCKHEED PV-1 (Mfr.) A manufacturer's photo shows a Navy PV-1 Ventura version of the Lockheed civil Model 18 Super Electra powered by two Pratt and Whitney R-2800 Double Wasp engines each of 2000 horsepower. Similar in general to the Air Force B-34 and military export versions, the PV-1 was a follow-on to the earlier Lockheed Hudson, but used mainly as a patrol plane, though the Marines did use a few modified for night fighting. They were really not suitable for that role. The first Navy PV-1s appeared in July of 1942.

LOCKHEED PV-1 (A.DiStasi) A PV-1 of an Atlantic anti-submarine warfare development squadron is shown in a wintry photograph. Normal armament was two fixed .50 caliber nose guns, two in a dorsal power turret, and two .30 caliber guns operated from the ventral location shown. A variety of stores could be carried in a fuselage bomb bay shown partly opened in the photo. Stores such as drop tanks or rockets or a searchlight could be carried underwing with the latter two being shown.

LOCKHEED PV-1 (Mfr.) Some of the 1600 PV-1 Ventura aircraft manufactured into 1944 had a solid nose equipped with a search radar; the two .50 caliber fixed guns are shown just above the radome in this photo. Additional fixed guns could be carried in the nose under the radome. The PV-1 had the Model 18 wingspan of 65 feet six inches and a length of almost 52 feet. Gross weight was 14-15 tons. The plane in the photo illustrates the wing drop tank installation.

LOCKHEED PV-1 (Mfr.) A dramatic photo of a 300-plus mph high speed PV-1 Ventura illustrates the twin gun dorsal turret, the astrodome, and antennas atop the fuselage. Also noteable are the sharply tapered wings with fixed outboard leading edge letter-box wing slots. Venturas operated in both the Atlantic and Pacific areas in World War II; a large number were provided to Australia and New Zealand.

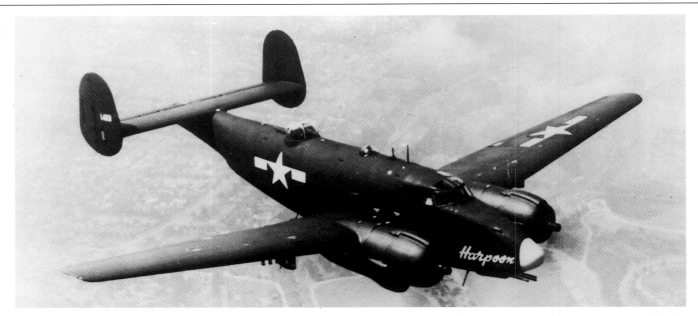

LOCKHEED PV-2 (Mfr.) A later version of the PV-1 Ventura, the PV-2 Harpoon was changed in having new wings and empennage on essentially the same fuselage. First appearing in early 1944 after a mid-1943 order for a batch of 500 aircraft, the PV-2 armament consisted of five .50 caliber fixed nose guns along with the twin gun dorsal turret and ventral guns. The revised shape of the vertical tails is noteable. With the same 2000 horsepower Double Wasp engines the heavier 18 ton gross weight Harpoon had a high speed of about 280 mph.

LOCKHEED PV-2 (P.Bowers) A post-war photo of a PV-2 converted for water bombing shows wing store stations still in place. Addition of 35 PV-2D aircraft in 1945 completed Harpoon production after many had been cancelled; these -2D models had a total of no less than eight fixed nose guns. A few PV-2C models were used for training. The PV-2 was used by the Naval Reserves until the late 1940s.

LOCKHEED XP2V-1 (J.Weathers) One of two prototypes of a long-lived patrol plane type, the seven place twin engine Lockheed XP2V-1 Neptune was the first new from-scratch design patrol landplane for the Navy. Initiated during World War II, first prototype flight took place just before war's end in May of 1945. A latter-day photo of 1977 shows one of the prototypes at the Naval Aviation Museum in Pensacola, Fl. The two Wright R-3350 engines of 2300 horsepower each gave the 29 ton plane a high speed of just under 300 mph. A total of 166 production P2V-1s followed in 1946 and became operational the next year.

LOCKHEED P2V-2 (Mfr.) The cleanliness of the early P2V- models is well illustrated in this manufacturer's photo. Only stand-offs for eight rockets under each wing and a belly radome interrupt the sleek lines. A fuselage bomb bay carried depth charges, bombs, or two torpedoes. Fifty caliber and 20 millimeter guns comprised the armament, including nose .50s and two in a dorsal turret with most -2s having a twin 20mm gun tail turret. High speed with the two Wright engines was about 300 mph. Gross weight climbed to over 30 tons. Wingspan was 100 feet and length 78 feet. More than 70 were built.

LOCKHEED P2V-3 (USN) A Navy photo of August 1949 shows a P2V-3 Neptune version, the first 30 being procured with Fiscal Year 1947 funds. Ports for three of the six nose guns can be seen. This version first joined the Navy a year earlier. A total of 53 -3 aircraft were procured using FY 1948 money and 30 additional P2V-3W versions were bought for the early warning mission with extra radar under FY 1949 funds. Water-injected Wright R-3350s with combat power of 3200 horsepower each increased dash speed to over 330 mph. P2V-3Z models were special transports and a few P2V-3Cs were converted for carrier takeoffs carrying nuclear weapons to give the Navy a nuclear strike capability.

LOCKHEED P2V-4/P-2D (R.Besecker) Fifty-two aircraft of the P2V-4 Neptune version were also obtained with FY 1949 funds with deliveriies late that year. As shown in this May 1962 photo at NAS Willow Grove, Pa. some were still around in reserve squadrons that year, and were redesignated as P-2Ds in the newly revised system. The photo illustrates the nacelle of a new Wright Turbo Compound engine with three power recovery turbines, the enlarged belly radome, and wingtip fuel tanks. High dash speed at "wet" engine power went up to about 350 mph.

LOCKHEED P2V-4/P-2D (AAHS) Another P2V-4 photo of an aircraft in the Oakland, Ca. Reserves shows the large size of the radome for the belly search radar and the wingtip fuel tank. Total fuel capacity was over 4200 gallons translating into well over 12.5 tons of gasoline. Patrol range was well over 4000 miles.

LOCKHEED P2V-5/DP-2E (USN) One of well over 400 new nine place model P2V-5 Neptunes is shown in flight, this aircraft acting as a director plane of Navy Squadron VU-3 with a jet powered target vehicle hanging underwing. This -5 model has been retrofitted with two Westinghouse J34 engines to add 3200 pounds of thrust from each jet to the power of two Wright 3250 horsepower Duplex Cyclones. First -5 flight took place at the end of 1950 after an initial allocation of FY 1950 funds for 45 aircraft. The -5 shown has had the original nose turret removed and a MAD gear installation has replaced the two 20mm tail guns.

LOCKHEED P2V-5F/P-2E (Mfr.) A P2V-5F, in 1962 changed to P-2E, of Navy Squadron VU-3 shows the large belly radome, glassed-in nose, and the extended tail with Magnetic Anomoly Detection gear incorporated. Gun turrets are also removed in this version, and the two Westinghouse J34 jet engines are added, this work done in the mid-1950s. Gross weight could go as high as 40 tons; with all engines operating this Neptune could almost touch a 400 mph dash speed. One of the large wingtip fuel tanks had a searchlight in the nose.

LOCKHEED P2V-5FD/DP-2E (R.Besecker) Another director aircraft for launching a jet-powered aerial drone target is shown in May 1965 with engine cowling unlatched to show a Wright R-3350 piston engine. The long tail extension for MAD gear is shown, though that equipment would not be used in a target director mission.

LOCKHEED P2V-5FE/EP-2E (R.Besecker) This P2V-5 aircraft was an anti-submarine warfare (ASW) version with both active and passive submarine detection systems included in a retrofit. The photo shows the glassed-in nose, raised canopy, belly radomes, and tail MAD gear as well as one of the large tip-mounted fuel tanks with stabilizing fin. Empty wing store stations are located outboard of the J34 jet engine pod. Range with a typical weapons load was about 3000 miles.

LOCKHEED SP-2E (P.Bowers) A fine photo of a Neptune configured for ASW work and using the Julie/Jezebel gear is shown with the smaller of two types of wingtip fuel tanks. The port tip tank incorporates a searchlight in the tank, the direction of which could be controlled by an observer in the nose. Propellers are Hamilton Standard types. Many Neptunes were provided to friendly countries under mutual aid treaties.

LOCKHEED P2V-6/P-2F (USN) A P2V-6 (later in 1962 a P-2F) is shown in a July 30, 1962 Navy flight photo. This -6 aircraft is unusual in still having the nose and tail 20mm gun turrets, less the cannon, no wingtip fuel tanks, and no underwing jet engine pods. A total of 83 P2V-6 Neptunes were procured with the initial aircraft flying in October of 1952.

LOCKHEED P2V-6T/TP-2F (R.Besecker) A photo taken in December 1962 shows a training version of the Neptune used by the Naval Reserves. The aircraft has weapons removed from nose and tail turrets and tip fuel tanks and jet engine pods removed from the wings. The small belly radome of the -6 version is retained. The Wright Duplex Cyclone engines developed 3500 horsepower maximum.

LOCKHEED SP-2H (J.Weathers) The photo shows an SP-2H of the Memphis Naval Reserves, Squadron VP-67, on November of 1976, well after the P-3 Orion had replaced the P-2 in front line squadrons. The slightly nose-down thrust line of the Hamilton Standard propellers shows up; this was done to minimize the propeller vibration excitation factor Aq under certain flight conditions.

LOCKHEED SP-2H (J.Weathers) The Neptunes hung in for a long while with the Reserves; another view of a Memphis Reserves SP-2H of VP-67 in November, 1976 is shown, again in standard ASW configuration. The SP-2H had a wingspan of 103 feet ten inches, a length of 91 feet four inches, and a gross weight of 37 to 40 tons. With only piston engines operating high speed was about 320 mph; with the thrust of the jet engines added speed rose to about 400 mph. The low cruising fuel consumption figures of the Wright Turbo Compound piston engines with their three integral power recovery turbines provided good long range performance.

LOCKHEED P2V-7S/SP-2H (R.Besecker) The -7 Neptune was the final version of the famous aircraft which first flew in early 1954. The -7 included the latest in ASW detection gear and the jet engines underwing were a standard item. Many sub-variants were produced and a large number were exported. The example shown had its picture taken at Travis AFB in August of 1970 while operating with the Naval Reserves. It is in the basic -7 configuration.

LOCKHEED SP-2H (R.Besecker) A reserves SP-2H (born P2V-7S) of Squadron VP-17 is shown in March of 1965. The Neptune started retiring about this time with the last going to the Reserves in 1970. They served there and in miscellaneous duties right through the 1970s. Doors of the fuselage stores bay are open just below the belly radome in the picture.

LOCKHEED NP-2H (R.Besecker) A Neptune used for special test purposes is shown at a 1970s airshow. The letter N was used to designate such special purpose aircraft like that used at the Naval Air Test Center, undoubtedly NATC Patuxent River, Md. A few Neptunes were also used for special operations, particularly in Vietnam. There were even heavily armed attack versions.

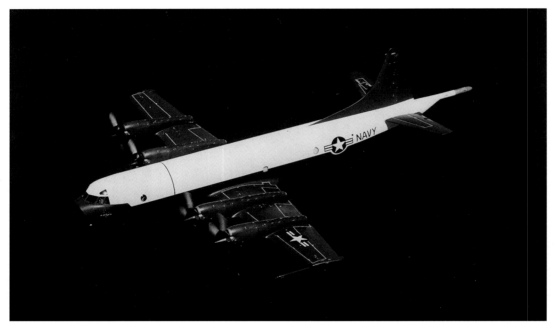

LOCKHEED YP3V-1 (USN) In April 1958 Lockheed won a competition for a Navy land-based ASW aircraft to replace the P-2 Neptune by proposing a navalized version of their civil Electra four engine turbo-prop transport, this to be called the XP3V-1 Orion. Shown is one of two YP3V-1 (later YP-3A) prototypes, the two first flying in August and late November of 1959. The long tail houses the MAD gear which had to be located as far away from the large metal mass of the engines as possible. The engines are Allison T56 turbo-props.

LOCKHEED P3V-1/P-3A (F.Dean) The initial production version of the ASW Orion was the P-3A using 4500 equivalent shaft horsepower Allison T56A-10W engines and flying first on April 15,1961. Deliveries began in the spring of 1962. The ten place aircraft shown here in the 1980s is from a reserve squadron at NAS Willow Grove,Pa. and a very cold photographer took the picture in early January. The searchlight under the starboard wing is clearly shown. The P-3A could carry torpedoes or a mix of torpedoes, mines, and depth charges in a fuselage bay or at underwing store stations.

LOCKHEED P-3A (USN) A fine photo of a P-3A Orion taken on May 18, 1965 shows the almost windowless pressurized cabin and radome nose. The Allison turboprops gave a total of 18000 installed shaft horsepower providing the 64 ton gross weight P-3A with a high speed of over 450 mph. Wingspan was 99 feet eight inches. A total of 157 P-3As were procured. The last 47 had a new Deltic ASW detection system installed and improved tactical displays.

LOCKHEED P-3B (R.Besecker) A P-3B Orion of Squadron VP-19 is shown at Moffat Field in California in December of 1972, about ten years after first flight of the type. Excluding those few exported to Australia, New Zealand, and Norway in 1966-69, the US Navy received a total of about 130 P-3Bs. New uprated Allisons of 4900 shaft horsepower each without water injection provided power to give the 70 ton gross weight plane a high speed of over 470 mph. Bullpup missiles could be carried underwing as well.

LOCKHEED P-3C (N.Taylor) A more advanced P-3C version of the Orion from Squadron VP-48 is shown in August of 1975, and illustrates the port underwing store stations. The P-3C got A-NEW system of updated sensors and control equipment. First flight was in September 1968, and service entry occurred in late 1969. The P-3C electronics have had continual updates through the years. Mines, depth bombs (including nuclear) and torpedoes in many combinations can be carried in a large fuselage bomb bay. A total of ten underwing store stations can carry additional weapons. High speed at a weight of 105000 pounds is 473 mph at 15000 feet altitude. Patrol speed is typically about 237 mph. at 1500 feet.

CONSOLIDATED XPY-1 (Mfr.) In early 1928 Consolidated won a Navy contract for a long range monoplane patrol flying boat with a crew of five in open cockpits designated the XPY-1 Admiral. The contract was for $150,000! The aircraft was designed to be convertable to a civil 32 passenger plane also and that was called the Commodore. Powered by two 425 horsepower Pratt and Whitney Wasp engines the XPY-1 wingspan was 100 feet. First flight was January 10, 1929 and first demonstration to the Navy was 12 days later on the Anacostia River with the Secretary of the Navy aboard. The single XPY-1 is shown in a company photo of February 7, 1929. Gross weight was 14000 to 16000 pounds.

CONSOLIDATED XPY-1 (USN) A three engine version of the XPY-1 Admiral is shown in an August 1929 test flight. Designer Mac Laddon called it a real monstrosity. The Navy wanted a high speed of at least 135 mph; the twin engine XPY-1 did less than 120 mph. The three engine version was too high to allow hangaring and servicing the high engine at sea was a real problem.

CONSOLIDATED XP2Y-1 (USN) The new patrol flying boat from Consolidated, the XP2Y-1 Ranger was contracted for in late May of 1931. Built in Buffalo, NY, the new Ranger was flown to Washington DC in April of 1932. Experiences in trying to test flying boats around Buffalo in winter started Consolidated on a search for a warmer climate. Again, as shown in the photo, the XP2Y-1 was tested with a third engine; it was no more successful than with the XPY-1. A big difference in the Ranger was use of a sesquiplane layout with a short lower wing supporting outboard floats.

CONSOLIDATED P2Y-1 (USN) The photo shows one of 23 production P2Y-1 Rangers. The initial aircraft was delivered to the Navy at Norfolk, Va. on February 1, 1933. These aircraft made many very long range pioneering flights in Navy service, one of which was a six plane flight from the US west coast out to Hawaii. The aircraft was powered by two Wright R-1820 Cyclones each of 575 horsepower giving the Ranger a high speed of about 125 mph. Gross weight was 21000 pounds; wingspan was 100 feet.

CONSOLIDATED P2Y-2 (USN) A photo of August 2, 1935 shows the new version of the P2Y- Ranger flying boat. After testing the final P2Y-1 modified into an XP2Y-2 with engine nacelles housing 750 horsepower Wright Cyclones in NACA long chord cowlings integrated into the wing leading edge in a position of minimum drag, the P2Y-1 planes were field-modified to that arrangement. Flight tests showed a 10 mph increase in cruise speed. Cleanliness was improved, but there were still a lot of struts!

CONSOLIDATED P2Y-3 (USN) With the successful testing of the XP2Y-2 an order was placed with Consolidated for 23 aircraft of a new model P2Y-3 Ranger on December 27, 1933. These aircraft were the last of the sesquiplane design with all the bracing struts. A few additional aircraft were exported, one each to Columbia and Japan and six to Argentina. Note the use of Curtiss Electric constant speed and variable selective pitch propellers which were excellent for use in maneuvering on the water.

CONSOLIDATED P2Y-3 (H.Thorell) The same aircraft as in the previous photo is shown on the water. The P2Y-2 updated and the P2Y-3s served the Navy well for about six years in front line service. They were to be replaced by the new XP3Y-1 and then the PBY- production types. Just before World War II 41 of the 46 P2Y- boats were sent to Pensacola, Fl. for use as training planes for a few more years.

CONSOLIDATED XP3Y-1 (USN) The March 1935 photo shows the prototype for what could easily be called the most famous flying boat/amphibian of all time. Designer Mac Laddon took advantage of all the mistakes he had by his own admission made previously, and an exceptionally clean flying boat design resulted. Competitive with the Douglas XP3D-2, the XP3Y-1 was the result of an October 1933 contract for a Ranger successor. Delivered to the Navy in early 1935, flight testing started March 21 at Norfolk, Va. Testing was very successful; the designation was changed to XPBY-1, and this led to production PBY-1 contracts. Powered with two 825 horsepower Twin Wasps, the 104 foot span ten ton XP3Y-1 had a high speed of almost 170 mph.

CONSOLIDATED XP4Y-1 (Mfr.) The XP4Y-1 Corregidor flying boat was an outgrowth of the Consolidated civil Model 31, a design started in July, 1938. Using a new efficient "Davis Wing" like the B-24 and two new Wright R-3350 2000 horsepower engines on a high cantilever monoplane wing along with twin tails, the XP4Y-1 also had a new design deep section hull as shown in the August 1943 photo. Another feature was use of integral beaching gear. First flight of Model 31 was on May 5, 1939. In February 1942 the Navy procured the Model 31 and supervised modifications to make it a naval patrol plane competitive with the Boeing XPBB-1.

CONSOLIDATED XP4Y-1 (Mfr.) A flight photo of the XP4Y-1 shows the navalized armed prototype. The nose gunner operated a 37mm cannon; there were twin .50 caliber gun turrets in dorsal and tail positions. Up to two tons of bombs could be carried. The two Wright Duplex Cyclones powered the 110 foot span 23 ton gross weight XP4Y-1 Corregidor to a high speed of almost 250 mph. Like the XPBB-1 the XP4Y-1 never got into production because flying boats were going out of favor and the R-3350 engines were needed for the Army B-29 bomber program.

CONVAIR P4Y-2G (P.Bowers) A postwar conversion of the Navy PB4Y-2 Privateer, the P4Y-2 was so redesignated in 1951 when the B for bomber in the original designation became irrelevant. Over 700 Privateers had been built, practically all in late World War II, and the Navy used them for a long time post-war. A few were provided to the US Coast Guard; one of these is shown in the photo. The plane was powered by four Pratt and Whitney Twin Wasp engines each of 1350 horsepower. With a gross weight of near 30 tons high speed was about 250 mph.

CONSOLIDATED P4Y-2G (P.Bowers) The photo shows an ex-US Coast Guard P4Y-2G after being sold for civilian use. The Coast Guard had removed all weapons stations and turrets and replaced these with large transparency areas for easy sighting during over water coastal patrols. The civilian use and the nature of the device along the trailing edge is not known.

CONSOLIDATED P4Y-2K (P.Bowers) Another version of the wartime PB4Y-2 Privateer after its redesignation to P4Y-2 in 1951 is shown. Among the postwar conversions were P4Y-2S ASW patrol planes, P4Y-2B stand-off bombers equipped with underwing radar guided glide bombs, and as shown in the photo P4Y-2K target drone aircraft, the latter becoming the QP-4B model in 1962 after the new redesignation system came into effect. Thus ended the career of the Navy Privateer aircraft.

CONSOLIDATED XP5Y-1 (Mfr.) One of two prototype four turboprop engine flying boats is shown during early taxi tests. In mid-1946 Consolidated, now Convair, obtained a contract for a new patrol plane with a state of the art hull design and four Allison XT-40 twin power section turboprop engines developing over 5000 horsepower each and driving dual rotation propellers, a very complicated powerplant arrangement. First flight took place in April, 1950 and the second aircraft never flew. The Allison turboprop engine posed many problems.

CONSOLIDATED XP5Y-1 (Mfr.) Graceful lines of the XP5Y-1 patrol boat prototype are shown in this flight photo. Wingspan was about 147 feet, gross weight about 65 tons, and high speed almost 400 mph at altitude. The XP5Y-1 had a planned defensive armament of no less than ten 20mm cannon in remotely controlled turrets. A variety of offensive mines, bombs, depth charges, and torpedoes could be carried. The aircraft was a failure primarily due to powerplant development problems; a 1953 crash ended the patrol plane program, but the Navy kept on with a transport version known as the R3Y-1.

NAVY PATROL BOMBER AIRPLANES

The Navy patrol bomber category lasted approximately from the mid-1930s through World War II and a little after. It started in 1936 when the Consolidated XP3Y-1 patrol plane prototype became the PBY-1 patrol bomber with bombing added as a secondary mission at a time when the Navy was favoring dual mission aircraft. The time of the patrol bomber ended not long after the war when it became obvious how vulnerable patrol planes could be on bombing missions and when the primary mission of the late VPB patrol bombers was really ASW. Actually the current P-3 Orion patrol aircraft can carry bombs and depth charges and torpedoes at least as easily as the old PBY- Catalina patrol bomber, so differences between VPB and VP airplanes have narrowed anyway. Ten different types were used over the years of the patrol bomber, six of them flying boats or amphibians, one a twin float seaplane, and three landplane conversions of Army bomber or attack airplanes. Of the ten PB (or PTB) types four stayed as single prototypes with no production leaving a total of only six Navy production patrol bomber aircraft.

The first patrol bomber so designated was Consolidated's twin engined PBY- monoplane of 1936, later named the Catalina and doubtless the most famous and most produced flying boat or amphibian of all time. During World War II versions of the Catalina were produced by the Naval Aircraft Factory as the PBN-1 and by Boeing-Canada as the PB2B-1 and PB2B-2.

In 1937 the first four engined flying boat in the VPB class appeared in the form of the single Sikorsky XPBS-1 "Flying Dreadnaught". This aircraft lost out to the Consolidated XPB2Y-1 Coronado for production. The same year an unusual triple mission single prototype aircraft was flown, the twin engined Hall Aluminum XPTBH-2 patrol torpedo bomber, unusual in the fact it was a large twin float seaplane. There was no production.

A year later Consolidated put out a four engined prototype flying boat patrol bomber competitive with the Sikorsky XPBS-1, the XPB2Y-1 which after considerable development and modification into a PB2Y-2 version became a production item used in World War II, but mainly as a transport, not as a bomber, nor even a patrol plane to any extent.

Martin in 1940 came out with their first patrol bomber, the PBM- Mariner flying boat with two engines and a deep hull. Put into production, it served through the war and a number of amphibious versions were produced after the war.

The Lockheed PBO-1 of 1941 was the famous Hudson bomber in naval form and the first landplane patrol bomber. A few served the Navy in ASW roles primarily using two Wright Cyclone engines.

In 1942 Boeing produced a new twin engined flying boat patrol bomber, the XPBB-1 Sea Ranger, later dubbed the "Lone Ranger" because no more were built. The planned production of the Sea Ranger was upstaged by the requirements of the Army B-29 bomber program.

Martin first flew their very large prototype Mars flying boat with four engines in 1942, however the aircraft was to prove more valuable as a transport plane, the PB2M-1R, and this was the prototype for the few JRM-1 utility transports so the Mars was never used as a patrol bomber.

In the middle of World War II the Navy found the need for a true bomber as the next patrol bomber and acquired some Army B-24 Liberator bombers as Navy Consolidated PB4Y-1s. The Navy thought so much of this airplane they devised a major modification of the Liberator with a single tail and new fuselage, the PB4Y-2 Privateer, later in post war days re-designated the P4Y-2.

The other Army bomber utilized by the Navy was acquired by them for the US Marine Corps. This was the PBJ- medium bomber, really an Army B-25. Both gun and glassed in nose types were used in later wartime as the PBJ-1H and PBJ-1J.

BOEING XPBB-1 (Mfr.) A very large deep-hulled twin engined patrol bomber design by Boeing with a wing like the B-29, the XPBB-1 Sea Ranger was ordered in mid-1940 and a single prototype first flew two years later. A special plant was built at Renton, Wa. to build 500 heavily armed Sea Rangers, but the facilities and the engines used were needed by the B-29 bomber program. There was thus no Sea Ranger production. Powered by two 2000 horsepower Wright R-3350 Duplex Cyclone engines, high speed of the XPBB-1 was 219 mph. Wingspan was 139 feet, eight inches; normal gross weight was 31 tons which could go to 50 tons using JATO.

BOEING PB2B-1 (Mfr.) A variant of the Consolidated PBY- flying boat, the PB2B-1 Catalina was the equivalent of the PBY-5, and was built by Boeing, Vancouver, BC. for the British Royal Air Force, and the photo shows a PB2B-1 in RAF markings. Production was started in mid-1943 and a total of 240 boats were turned out. Some went to Australia and New Zealand. Characteristics were similar to those of the PBY-5.

BOEING PB2B-2 (Mfr.) A later version of the Boeing Canada built Catalina was similar to the PBY-6 and PBN-1 and called the Catalina VI by the British. Sixty-seven were built in 1944-45, most of which were exported to Australia. Wingspan was 104 feet three inches, gross weight 37000 pounds, and two 1200 horsepower Pratt and Whitney Twin Wasps gave the PB2B-2 a top speed of 175 mph and a cruising speed of 117 mph. Armament was three .50 caliber and one .30 caliber guns; bomb load was two tons. Range was about 2500 miles.

NORTH AMERICAN PBJ-1H (USN) The PBJ-1 designation was used by the Navy to denote Army B-25 twin engine Mitchell bombers used by the Marine Corps in World War II. The suffix letter of the PBJ- series corresponded to that of the Army model; thus the PBJ-1H shown aboard a carrier was a B-25H with a solid nose and very heavy fixed nose armament. The -1H was not normally used aboard a carrier; a total of 248 were made for Navy/Marine use late in the war. Powered by two Wright R-2600 engines each of 1700 horsepower, the 17 ton gross weight aircraft had a high speed of about 275 mph.

NORTH AMERICAN PBJ-1J (USN) Another version of the Mitchell bomber used by the Marines was the PBJ-1J. A PBJ-1J of the Navy is shown under test at NATC Patuxent River, Md. on May 31, 1944, one of 255 delivered. Note the side package guns and the belly turret on the aircraft. Installed power, speeds and weights of this version were similar to those of the PBJ-1H. Over 200 earlier model Mitchells were also procured by the Navy.

MARTIN XPBM-1 (Mfr.) After tests of a small scale two place model in 1937 Martin went ahead with construction of a prototype XPBM-1 full size twin engine patrol bomber ordered by the Navy in that year. The prototype is shown under test in a company photo of February 24, 1939 soon after initial launch. This single aircraft was distinguished by a gulled wing of 118 foot span, a flat horizontal tail with twin vertical tails and retractable tip floats. Two Wright R-2600 engines of 1600 horsepower each powered the 20 ton gross weight deep hulled boat to a maximum speed of about 210 mph.

MARTIN PBM-1 (USN) Two PBM-1 Mariners lead two PBM-3s in a Navy formation photo. The initial aircraft of 20 PBM-1s produced with a crew of seven, it was delivered to Navy Patrol Squadron VP-55 on September 1, 1940. It had retractable tip floats and a dihedraled horizontal tail. Nacelle bays could carry up to a ton of bombs. The bow turret shown had a .30 caliber gun; a dorsal turret and two fuselage side positions had .50 caliber guns.

MARTIN PBM-3 (USN) An early PBM-3 Mariner patrol bomber is pictured in a Navy photo. These early aircraft were utilized in training. They differed from the -1s in using longer fixed instead of retractable wing floats supported by struts and by larger engine nacelles housing uprated Wright R-2600 engines each of 1700 horsepower. As can be seen, the fuselage side gun stations were also modified. The twin vertical tails were mounted normal to the horizontal surfaces so they canted inward.

MARTIN PBM-3 (USN) A Navy flight view of the PBM-3 Mariner taken near the Banana River in Florida in March, 1943 shows the plan view of the big boat with the high aspect ratio wing and dorsal gun turret apparent. The original -3 contract called for 379 aircraft; several different versions resulted.

MARTIN PBM-3C (USN) A PBM-3C reconnaissance bomber variant is shown running up while on its beaching gear. The -3Cs, 272 of which were built, went into Navy service in September of 1942. Armament was increased with use of twin .50 caliber guns in a nose power turret as shown in the photo and single .50s in dorsal, tail, and two side positions. The big radome forward on the fuselage is for a search radar. The two R-2600 engines gave the 26 to 29 ton gross weight boat a high speed of just about 200 mph. Wingspan was 118 feet and length 80 feet.

MARTIN PBM-5 (USN) After over 675 PBM-3 Mariners in various versions had been delivered by Martin production shifted in mid-1944 to a PBM-5 version, one of which is shown in a JATO (jet-assisted takeoff, more properly called rocket-assisted takeoff) with auxiliary solid fuel rockets mounted at the side gun stations. The PBM-5 changed to use of two 2100 horsepower Pratt and Whitney Double Wasp engines giving a higher maximum speed of over 210 mph. About 600 were produced, some even post-war.

MARTIN PBM-5A (Mfr.) Three dozen PBM-5A amphibian versions of the Mariner were produced in 1948-49 using, as shown in this company takeoff photo, a tricycle landing gear arrangement. The Double Wasp engines drove four bladed propellers. These aircraft were utilized by the US Coast Guard as replacements for their PBM-5G boats.

MARTIN PBM-5G (R.Stuckey) Mariners were produced even after World War II was over. The US Coast Guard used a few as PBM-5Gs with these stripped of armament and air-sea rescue sighting stations replacing gun positions. A good view of the main and tail beaching gear elements is shown in the photo; it is easy to see why an amphibian version was desireable. Four blade Curtiss Electric propellers provided good thrust control for water maneuvering.

MARTIN PBM-5G (P.Bowers) Another view of a beached PBM-5G Mariner of the Coast Guard, this one with a newer APS-31 radar installation on the fuselage just aft of the pilots' cockpit area. The Coast Guard Mariners and similar Navy types served well into the 1950s.

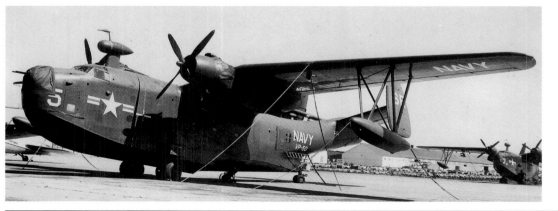

MARTIN PBM-5S2 (R. Besecker) A PBM-5S2 update of some wartime PBM-5s is shown in a post-war photo of aircraft number five in Navy Patrol Squadron VP-50. Intended for ASW missions, the -5S had search radar for submarine detection and some armament was retained. Note the nose guns and the open doors of the weapons bay in the engine nacelle aft section.

MARTIN PBM-5S2 (USN) As major production aircraft of World War II, Martin PBM- flying boats, with over 1300 produced, were used in many post-war roles including special test work. This Navy photo illustrates testing of a hydroski installation on a PBM-5S2 in the 1960s. Another Mariner was used in naval tests of the so-called tilt-float principle, this being employment of long slim floats supporting an ASW aircraft above rough water without undue motion and using dipping sonar.

MARTIN XPB2M-1 (Mfr.) A single four engined aircraft ordered in August, 1938, the very large (200 foot span; 117 foot three inch length) Martin XPB2M-1 Mars Navy patrol bomber with a crew of eleven was launched on November 8, 1941, just before America's entry into World War II. The big boat is shown at launching adorned with flags in the company photo. Test problems delayed first flight until July of 1942. Powered by Wright R-3350 Duplex Cyclone engines each of 2200 horsepower the 72 ton Mars had a high speed of about 220 mph. No more of these patrol bombers were ordered.

MARTIN XPB2M-1R (Mfr.) There was no requirement for the Mars as a patrol bomber in World War II and in 1943 the single XPB2M-1 prototype was modified to an XPB2M-1R cargo transport by floor reinforcement and hatch enlargement, the modification first flying in December, 1943. The next year this Mars transport made several record flights, including one to Hawaii with a cargo load of over 10 tons. The Navy ordered 20 more Mars aircraft as JRM-1 transports, but after the war the order was reduced to five planes produced as JRM-s.

LOCKHEED PBO-1 (USN) One of 20 Lockheed Hudsons designated PBO-1 in the Navy, and actually Army A-29s needed for Naval ASW patrol is shown. Delivered in late 1941 and assigned to Navy Squadron VP-82 as the initial patrol landplane design, they were used in Newfoundland and achieved success shortly by sinking two German submarines in March of 1942. Powered by two Wright 1200 horsepower Cyclones, the ten ton gross weight Hudson had a top speed of about 260 mph. These aircraft had two fixed nose guns and dorsal and ventral flexible guns. They could carry 1300 pounds of depth charges.

SIKORSKY XPBS-1 (Mfr.) A competitor with Consolidated to build a large four engine flying boat, the XPBS-1 Flying Dreadnaught design obtained a Navy contract on June 25, 1935. First flight of the single prototype took place in August, 1937 with Navy delivery October 12, 1937. Though performance bettered that of the Consolidated PB2Y-1 Coronado, the XPBS-1 lost out on price, so only the prototype was built, but the aircraft flew many Navy missions and accumulated 1366 hours of flight service. It hit a log on landing at Alameda, Ca. and was badly damaged on January 30, 1942. Powered by four 1050 horsepower Pratt and Whitney R-1830 Twin Wasp engines, the 24 ton gross weight XPBS-1 had a cruising speed of 168 mph and a high speed of about 225 mph.

SIKORSKY XPBS-1 (Mfr.) The XPBS-1 is shown taxiing during flight test. Wingspan was 124 feet and length 76 feet. The big boat carried a .50 caliber gun in front and in a rear turret and .30 caliber guns at fuselage side positions. Nose and tail manual turrets can be seen in the photo. Though no more Navy models were ordered, three civil VS-44A boats based on the same design were ordered for American Export Airline. Water handling characteristics of the design were very good.

CONSOLIDATED PBY-1 (Mfr.) The flight tests of the Consolidated XP3Y-1 patrol plane were very successful as were those of the competitive Douglas XP3D-1 and -2, and the choice came down to one of price. Consolidated bid $10,000. lower per plane ($90,000.) and won. The photo shows a PBY-1 on September 22, 1936, either the first or one of the first production aircraft of 60. Navy Squadron VP-11 got the first PBY-1 on October 5, 1936. Using two Pratt and Whitney Twin Wasp 900 horsepower engines, maximum allowable gross weight was 25,236 pounds and high speed 184 mph at 8000 feet. A total of 1750 gallons of fuel gave a maximum range of about 4000 miles. There was a .30 caliber gun at bow and tunnel positions and a .50 caliber at each waist location. A bomb load of up to 2000 pounds could be carried.

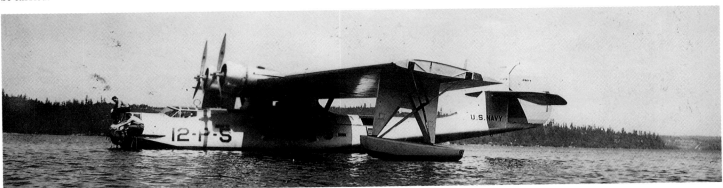

CONSOLIDATED PBY-1 (H.Thorell) A PBY-1 of Squadron VP-12 is shown on the water with two crewmen attending to the mooring. The boat was very clean compared to earlier designs with the 104 foot span wing mounted on a pedestal with four supporting struts and retractable wingtip floats. Landing speed of the -1 was 64 mph and service ceiling 22000 feet.

CONSOLIDATED PBY-2 (Curtiss) Before the first production delivery of a PBY-1 the Navy was so interested in the design they placed another production order for 50 PBY-2 aircraft at a cost of $4,898,000. with little more than minor changes in equipment. These were delivered to the Navy beginning in May, 1937. The photo shows a PBY-2 of VP-11 pictured on November 10, 1937.

CONSOLIDATED PBY-2 (Mfr.) Pleasing lines of the PBY-2 show up in this company photo of plane number 12 of Squadron VP-11. The big wing had a total area of 1400 square feet so at maximum gross weight of 28600 pounds the wing loading was still only a low 20 pounds per square foot. The vertical tail had been modified considerably from the prototype XP3Y-1. Engines were the same as on the -1 and they drove Curtiss propellers.

CONSOLIDATED PBY-2 (USN) A flight of PBY-2 patrol bombers provides an aesthetic view of the big boats which were used for many record-breaking long distance flights. They cruised at about 100 mph and had a service ceiling of about 21000 feet. Squadron VP-11 was the first to receive PBY-2s. The initial order for this version was on July 25, 1936.

CONSOLIDATED PBY-3 (Mfr.) A new PBY-3 as yet unassigned to a Navy squadron is pictured by a company photographer flying along the California coast. A total of 66 PBY-3s with uprated 1050 horsepower Pratt and Whitney Twin Wasp engines were ordered in late 1936. The -3 was the last production version to have all sliding hatches at the fuselage beam positions. Increased engine power upped high speed to about 190 mph and normal and maximum gross weights were 22700 and 29000 pounds respectively.

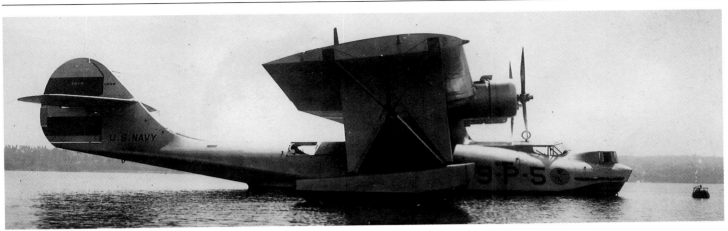

CONSOLIDATED PBY-3 (H.Thorell) A PBY-3 of Patrol Squadron VP-9 is shown at its mooring with aft hatches open. An ADF loop antenna is mounted just over the aft area of the cockpit and the front circular turret for a .30 caliber gun is just ahead of the cockpit enclosure. The PBY- aircraft had a maximum fuel capacity of 1750 gallons, and with this could fly up to a maximum range of well over 4000 miles. The PBY-3 came into service in 1938.

CONSOLIDATED PBY-4 (USN) A Navy flight photo shows a PBY-4 Catalina of Squadron VP-13. A total of 33 were procured in 1938-39. Little was changed, though the last three had modified blister waist gunner fairings and another got revised into an XPBY-5A amphibian flying initially on November 22, 1939. The retractable tricycle landing gear was successful though empty weight increased by 2300 pounds. PBY-4 deliveries were completed in mid-1939, and then there was a hiatus in PBY- Catalina production for 15 months.

CONSOLIDATED PBY-5 (Mfr.) A company photo shows a PBY-5A Catalina taking off. The picture shows clearly the retraction mechanism for the wingtip floats. On December 20, 1939 the Navy ordered 200 PBY-5s in what was the largest single Navy aircraft order since World War I. The first three were delivered in November, 1940 (along with three export Model 28-5s for Britain's RAF). They were distinguished by new blisters at the waist positions, a new version of the Twin Wasp engine rated at 1200 horsepower each , and a revised rudder shape. The Navy had accepted 167 PBY-5s as flying boats by September, 1941; the last 33 were produced as PBY-5A amphibians.

CONSOLIDATED XPBY-5A (USN) A Navy flight photo shows the XPBY-5A amphibian Catalina version, a modified PBY-4 with the old style sliding hatches, displaying its new retractable tricycle landing gear in the down position ready for landing. Even with the empty weight increase of over a ton the increase in operational flexibility of an amphibian was deemed worth the penalty.

CONSOLIDATED PBY-5A (Mfr.) One of the last of the 200 plane order, the -5A amphibian Catalina is depicted in a company photo (*above*). The new large blister-type gunner and observer station at the fuselage waist is shown along with the revised rudder shape. The name Catalina came from the British in late 1940. The location of the nose gear wheel well can be seen. The two 1200 horsepower Pratt and Whitney Twin Wasps gave the 34000 pound gross weight amphibian a high speed of about 180 mph.

CONSOLIDATED PBY-5A (AAHS) Another view of a PBY-5A Catalina taken later shows a radome for search radar above the forward fuselage. The generally clean lines of the aircraft show up with the wingtip floats retracted, there being no need for them down with the aircraft operated as a landplane. Some PBY-5As, 59 in all, were built in 1944-45 in a New Orleans plant originally intended for production of the aborted Consolidated Model 31/XP4Y-1. By December of 1941 two squadrons operated PBY-5As made by Consolidated.

CONSOLIDATED PBY-5A (AAHS) A PBY-5A used by the US Coast Guard for air-sea rescue is shown with the rescue lifeboat carried under the starboard wing and faired rather neatly into it. The search radar installation also is in place. In addition the USAAF operated PBY-5As, 54 of which were designated as OA-10 observation amphibians. Many PBY-5As were also operated by other countries, principally Britain's RAF, but also the Free French forces, Canada, and New Zealand.

CONSOLIDATED PBY-5A (Mfr.) A fine company flight view of a Catalina shows the broad wing of the aircraft and the two waist blisters. Considering all the production sources in the US, Canada, and Russia, there were an estimated 2029 flying boats and 1402 amphibians produced. These figures include the assumption that Russia produced 150 boats.

CONSOLIDATED PBY-6A (Mfr.) Flight view of the final Catalina version, the PBY-6A amphibian, shows the principal external difference as the new tall vertical tail giving greater rudder effectiveness. The -6As were built at New Orleans by Convair and 175 aircraft were produced from March to September of 1945, the end of the war. Of these 30 went to Russia. The other versions with a tall tail were the 155 PBN-1 Nomad flying boats built by the Naval Aircraft Factory from early 1943 through March 1945 with almost all going to Russia, and the 67 PB2B-2s mentioned previously.

CONSOLIDATED XPB2Y-1 (Mfr.) A new four engine prototype patrol bomber competitive with the Sikorsky XPBS-1 was the result of a May 26, 1936 contract award. The Navy supplied $600,000. and Consolidated $400,000. to design and fabricate the big flying boat. First flight was on December 17, 1937, and dangerous lateral-directional stability was apparent, particularly at low power settings. A company photo shows the aircraft after the first vertical tail fix of adding two fins. Later another tail with two vertical surfaces at the end of a dihedraled horizontal tail was added, this in 1938. Powered by Twin Wasp engines each of 1050 horsepower the 25 ton XPB2Y-1 had a top speed of 230 mph. It was used as a command transport by high level Navy officials.

CONSOLIDATED PB2Y-2 (Mfr.) A company photo of December 8,1940 shows one of six nine place PB2Y-2 Coronado patrol boats about three weeks before delivery to Navy Squadron VP-13. Changes from the prototype included larger twin tails, uprated 1200 horsepower Twin Wasp engines, a revised hull, and increased armament including six .50 caliber guns. Bombs are shown under the wings. Normal gross weight increased to about 30 tons and high speed to 255 mph. The large size of the new boat can be judged by the size of men in the photo. Wingspan was 115 feet.

CONSOLIDATED PB2Y-2 (Mfr.) Two more views of the first PB2Y-2 (*above, right*) show details of the production dihedraled twin tail with increased area. The .50 caliber gun positions were in the nose, tail, dorsal turret, two side, and in a ventral location. The wingtip float retraction system was similar to that of the earlier Catalina. One each of the six PB2Y-2s was converted into an XPB2Y-3 and XPB2Y-4, so there was little if any true operational use of the -2 models; most were involved in test work.

CONSOLIDATED PB2Y-3 (USN) A Navy photo shows the first real production version of the Coronado, the PB2Y-3, in a flight view that emphasized the increased hull depth. Armor and protected fuel tanks were featured along with three twin .50 caliber power gun turrets. A total of 210 aircraft were ordered in November, 1940 with the final plane delivered in October, 1943. A few Coronados went to the UK. Gross weight increased to 34 tons and high speed was about 225 mph. These aircraft were shortly to be replaced by more efficient landplanes for patrol bomber duties.

The big PB2Y Coronado type was the last basic design of a four engined flying boat for the Navy that entered production. Designing a boat hull for water operation entailed considerable aerodynamic penalty.

CONSOLIDATED PB2Y-3R (Mfr.) A total of 41 PB2Y-3 Coronado boats were converted to transports with ten of these going to the British. The transports were called PB2Y-3Rs or PB2Y-3Bs and had armament removed. One is shown in a company takeoff shot. Gross weight was about four tons less than the patrol bomber version. A PB2Y-3R could carry a crew of five and up to 44 passengers, or a mix of 24 passengers and 8600 pounds of cargo, or an eight ton cargo load.

CONSOLIDATED PB2Y-5 (Mfr.) A company flight view shows the final version of the Coronado, the PB2Y-5 equipped with lower altitude rated Twin Wasp engines like the Catalina and considerably increased fuel capacity to provide greater range. Nose and dorsal power turrets can be seen. The PB2Y-5s were -3s converted. Some had radomes added behind the cockpit area. A few -5s were made into -5H hospital planes with provisions for two dozen litter cases. The single XPB2Y-4 was a -3 using Wright R-2600 Double Cyclone engines.

CONSOLIDATED PB4Y-1 (USN) In mid-1942 after earlier Navy pressure for a share of B-24 Liberator production an agreement was reached and the first Navy B-24Ds were designated as PB4Y-1s. Eventually almost 1200 Liberators were Navy-designated as the war continued and later models went to to the Navy as they appeared into 1945. These aircraft had a patrol range covering large areas of ocean, and PB4Y-1s sank several submarines. A typical PB4Y-1 was powered by four 1200 horsepower Pratt and Whitney Twin Wasp engines, had a 110 foot wingspan, and a 30 ton gross weight. Top speed was about 285 mph; cruising speed for long range was about half that figure.

CONSOLIDATED XPB4Y-2 (USN) A Navy photo of June 6, 1944 shows one of three XPB4Y-2 Privateer prototypes converted from B-24D/PB4Y-1s with major modifications. First flight of a prototype took place in September, 1943. These Privateer prototypes resulted from the success of the -1 Liberators and the desire to completely navalize the design. Wings and landing gear were not changed; engines were revised as well as the fuselage and tail surfaces. Powered with four 1350 horsepower Twin Wasps lacking turbosuperchargers, the new aircraft had six gun turrets with eight .50 caliber guns and could carry over six tons of bombs.

CONSOLIDATED PB4Y-2 (R.Besecker) This view of a PB4Y-2 Privateer shows its gun armament well. A twin .50 caliber power turret is at the nose and two twin gun turrets are located atop the fuselage with only the barrels of the aft turret guns seen. A power turret is located in the tail and one of the twin gun positions at the fuselage waist is pictured. Bay doors for up to four tons of bombs are open and worked just like those of the Army B-24. The single tall tail provided better clearance for gunfire than the earlier twin tails of the PB4Y-1.

CONSOLIDATED PB4Y-2 (USN) A total of 740 PB4Y-2 Privateers were built after a production order of late 1943; the first deliveries began in March, 1944 and lasted just past the end of World War II. In wartime each aircraft normally carried a crew of 12. Radar, ECM, and extensive radio equipment was carried. Some Privateers continued in service well after the war; an example is shown in this Navy photo of September 9, 1946.

CONSOLIDATED PB4Y-2M (USN) This Navy photo of July 18, 1946 is of a PB4Y-2M, with the suffix letter meaning it was used for carrying early missiles underwing. This aircraft may have been one of those Privateers carrying anti-shipping missiles called Bats, a winged radar-homing type used very late in World War II. Gun positions have been eliminated or modified.

HALL XPTBH-2 (USN) A distinctly unusual design with an unusual designation, the Hall XPTBH-2 was a large (79 foot four inch wingspan) twin engine multi-mission twin float seaplane ordered in mid-1934 as an XPTBH-1 patrol torpedo bomber with Wright Cyclone engines. The single prototype appeared in early 1937 powered by two Pratt and Whitney Twin Wasps each of 800 horsepower driving Curtiss propellers. The fuselage had an enclosed bay for bombs or a torpedo and three machine guns were located in nose, dorsal and ventral positions. Gross weight was nine to ten and one half tons and high speed was about 185 mph. The Navy decided they had no interest and no production resulted.

NAVY TRANSPORT AIRPLANES

In 1930 transport airplanes were considered as an appropriate category of Naval Aviation, and a VR class was established. Types already in service in other categories that fit the transport definition were redesignated. These included Ford and Fokker trimotors and Sikorsky patrol boats. The Navy seldom if ever had the money allocated to develop specialized transports tailored just to their needs; most often naval transports were conversions of transports available on the commercial market.

In 1930 there were several Fokker trimotors in service as TA-types (T for transport) and these were redesignated as RA- types with the R now for transport leaving the T designator for torpedo planes. Similarly Ford trimotors in service like the XJR-1 purchased in 1927 as the Navy's first transport plane were redesignated from JR-s (J for utility) to RR-s. Again Sikorsky PS- aircraft (P for patrol) became RS- models.

In 1931 four civil aircraft were purchased as transport planes by the Navy, the first a single Curtiss Kingbird twin engined land high wing monoplane light transport designated an RC-1. It served the Marines at Quantico, VA, and at San Diego, Ca. until 1936. The second type was the first of a series of popular Douglas Dolphin civil twin engined high wing monoplane amphibians as the RD-series for Navy, Marines, and Coast Guard. The third type was the Wasp powered Bellanca civil high wing monoplane, the first of these serving as a radio test plane at NAS Anacostia, Md. as the XRE-1 to -3. The fourth airplane was a single civil Lockheed Altair single engine low wing monoplane as a VIP XRO-1 transport.

The year 1934 saw introduction in the Navy of two new twin engine transports, one a big Curtiss Condor biplane with two Wright Cyclones used by some airlines. In the Navy it was an R4C-1. A more modern type was the pioneering new Douglas DC-2 with two Wright Cyclones and all metal structure. The Navy had just a few as R2D-1s, but many civil models were operated by airlines.

In the next year, 1935, the Navy purchased for the Coast Guard a four place Stinson Reliant high wing monoplane as an RQ-1 light transport and a Northrop Delta all metal low wing monoplane for the same service as a VIP transport designated an RT-1.

Still in the depression era in 1936 there was little Navy money for transports but they managed to purchase three single engine Kinner Envoy civil types for high level officer transport as XRK-1s along with single examples of two different versions of the sleek new twin engined Lockheed civil Model 10 Electra all metal low wing monoplanes as the XR2O-1 and XR3O-1, the latter model for the Coast Guard.

The next Navy transport purchases were made in 1939 to sample the latest in civil transport designs including several navalized versions of the unusual modern Douglas DC-5 high wing all metal twin engine transport meant for but never used by the airlines; these were R3D-1 and R3D-2 models in the Navy and USMC respectively. In addition a single Lockheed Model 14 was obtained as an XR4O-1 for VIP transport use.

In late 1941 the US went to war, and the principal Navy transport, along with the Army, was a version of the civil twin engined low wing 21 passenger Douglas DC-3 monoplane, designated as the R4D- with many versions used through wartime and for many years after as one of the most famous airplanes of all time. A final version developed specifically for the Navy was the R4D-8 with many improvements. Another 1941 Naval transport was the R5O- version of the Lockheed civil model 18 Lodestar which also saw wide World War II use.

In 1942 the Navy was purchasing large R5C-1 twin engined monoplane transports from Curtiss as Marine Corps versions of the Army C-46A Commando cargo transport. At the same time long range land based transport needs of the Navy were met now with the Douglas R5D- four engined transport equivalent of the civil DC-4 and the Army C-54.

During the mid-war period of 1943-44 the Navy ordered an unusual RB-1 high wing tail loading twin engine cargo transport named the Conestoga using stainless steel structure and made by the Budd company of Philadelphia. The few built never saw active Naval service however. In addition in 1944 the Navy RY- transport was their version of the Army Consolidated B-24/C-87 Liberator bomber/transport. And an RY-3 was developed from the PB4Y-2 patrol bomber.

Final World War II Navy transport designs of 1945 were the R7V- Constellation, a version of the four engine civil Lockheed transport with some variants later used for turboprop engine testing, and the R2Y-1, a Convair prototype using a new pressurized transport fuselage with Liberator wings and engines. The Convair airplane remained just a prototype.

Just post-war in 1946 Lockheed produced strictly for the Navy a very large double deck pressurized four piston engined transport as the XR6O-1 Constitution, but only two were built and the airlines never used the type.

A specialized twin engined twin boom rear loading cargo transport used by the Army as the C-119 Packet from Fairchild was adapted for Navy/Marine cargo use in 1949-50. The R4Q-1 and R4Q-2 were equivalent respectively to the C-119B/C and the C-119F Army models.

To give the US Coast Guard a couple of new VIP airplanes the Navy obtained Martin 404 civil twin engined transports in 1951 as RM-1s. More importantly Douglas DC-6B four engine civil transports modified to Navy needs were procured in the early 1950s as the R6D-1 for operations with the Military Air Transport System for many years. Some were used as R6D-1Z VIP versions.

In 1954 the large four turboprop engined Convair flying boat originally developed as an XP5Y-1 patrol plane design was modified for strictly transport use as the R3Y-1 and R3Y-2 Tradewind models with the -2 featuring bow loading doors and jet fighter aerial refueling capabilities, however the same turboprop powerplant problems dogged the R3Y-s, and only a few of each were built.

The next year saw the advent of a long-lived and very successful land-based rear loading transport using four Allison T56 turboprop engines, the Lockheed C-130 Hercules utilized in the late 1950s in various Navy missions along with those of the Army, including Antarctic explorations.

Navy C-130s were for a time designated as GV-1 Marine tanker aircraft.

The mid-1950s saw the Navy use of Convair civil Model 340 and 440 twin piston engine transports for both cargo and passenger use. The 340s were R4Y-1s and the 440s R4Y-2s. They were later revised in the C-131 category for the 1962 redesignation of service

aircraft. A new transport role had come into prominence during the Korean conflict. The COD (carrier on-board delivery) type first performed with converted TBM-3Rs. To provide a more effective airplane for COD operations the Grumman S2F- design was revamped into a TF-1 in 1955, later a C-1A in 1962.

In the 1960s two other Navy transport aircraft came into use, the first another COD type as a replacement for the C-1A. In 1964 the new Grumman twin turboprop C-2A Greyhound appeared as a cargo modification of the E-2 early warning type with a new fuselage. These aircraft have been updated and still serve as the Navy's COD airplanes. Another type used by the Navy for specialized missions including training functions and for the Coast Guard is the Grumman C-4 versions of their civil Gulfstream twin turboprop transport plane. Use of these planes started in the 1960s.

A few Douglas DC-9 twin jet transports were purchased by the Navy in the early 1970s as C-9B Skytrains.

It is expected that future Navy transport needs will be filled by conversions of newer civil transport models or as additional updates of currently used types. No new specialized naval transport models have yet appeared on the horizon.

FOKKER RA-1 (USN) A cargo modification for Naval/Marine use of the Fokker commercial F-7A trimotor, the RA-1 type was a redesignation of the TA-1 because the T was being used for torpedo planes. The three RA-1s were employed by the US Marine Corps in Nicaragua in the late 1920s and the Marine insignia is shown on this ten place plane #2 in a Navy photo of November 23, 1929. The RA-1 had three Wright Whirlwind engines of 220 horsepower each giving the four and one half ton gross weight Fokker a high speed of about 115 mph. Wingspan was 63 feet five inches. These aircraft became RA-3s when re-engined with 300 horsepower Whirlwind engines.

FOKKER RA-2 (USN) Again originally designated as TA-2s, the three 11 place RA-2s had an increased wing area with a span of almost 73 feet. Powered with three Wright R-790 Whirlwinds each of 225 horsepower the 10400 pound gross weight RA-2/TA-2 Fokkers had a top speed of about 112 mph. The 1929 Navy photo of an RA-2 shows plane #5 used by the Marines. Note the external load tied under the fuselage. The RA-2s were later updated to use 300 horsepower Whirlwind engines and became RA-3s.

FOKKER RA-4 (USN) The final airplane of the Fokker RA- series was a navalized civil F-10A airliner. The single RA-4 trimotor for the Marines used three Pratt and Whitney Wasp engines each of 450 horsepower giving the six and one half ton gross weight transport a high speed near 150 mph. Wingspan was 79 feet three inches. The RA-4 was unsuccessful and no more of the type were ordered.

BUDD RB-1 (P.Bowers) An unusual aircraft because of its stainless steel construction by the Budd company of Philadelphia, Pa., the RB-1 transport was a twin engined tail loader with a rear ramp. A total of 26 planes were finished and 17 were delivered. The RB-1 contract was given Budd in August, 1942 after a letter of intent in April. Flight tests started October 31, 1943 and the first plane was delivered in March of 1944. The RB-1s never saw active service; they were bought surplus by Flying Tiger Airlines. Wingspan was 100 feet. Two Pratt and Whitney Twin Wasps each of 1050 horsepower gave the 33850 pound gross weight RB-1 a high speed of 197 mph.

CURTISS RC-1 (H.Thorell) In early 1931 the Navy took delivery for the Marines of a single Curtiss Kingbird civil light cabin transport monoplane. An unusual configuration with a short fuselage nose where the propeller discs were directly in front of the pilot's eyes, the twin engined twin tailed aircraft went first to squadron VF-9M and later to VJ-7M. Wingspan was 54 feet six inches. The two Wright Whirlwind engines each of 300 horsepower gave the 6100 pound gross weight Kingbird a top speed of about 135 mph.

CURTISS R4C-1 (USN) The Navy photo (*above*) illustrates a navalized version of a Curtiss Condor civil transport of 1934, two of which were procured as Marine utility transports in that year. Their final use was in a 1940 expedition to Antarctica. These big 82 foot span biplane transports were powered by two 700 horsepower Wright R-1820 Cyclone engines which provided a high speed capability of about 180 mph. Gross weight was 17500 pounds. It is interesting that R4C-1 biplanes with wings that could not be deiced would be sent to Antartica.

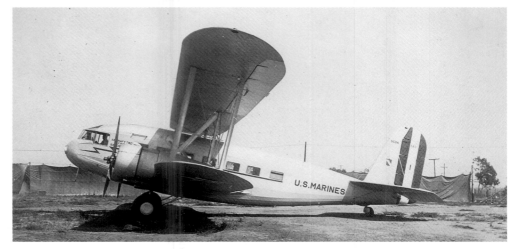

CURTISS R5C-1 (R.Stuckey) A post World War II photo shows one of 160 Curtiss C-46A Commando twin engined transports acquired by the Navy, mainly for the Marines in the Pacific during the war. The Commando was designated an R5C-1 in the Navy and USMC. Powered by 2000 horsepower Pratt and Whitney R-2800 Double Wasps, the 108 foot wingspan 28 ton R5C-1 had a high speed of about 265 mph.

DOUGLAS RD (AAHS) One of three Douglas Dolphins acquired by the US Coast Guard in 1931 is shown distinguished by the triple vertical tail fin arrangement and, like all Dolphins, by the twin Wright Whirlwind engine nacelles strut-braced above the monoplane wing and tied together by a small airfoiled "wing". The main wing is equipped with long ailerons but no flaps. These aircraft were used into the late 1930s.

DOUGLAS XRD-1 (H.Thorell) A Navy version of the popular Douglas Dolphin amphibian monoplane was procured early in 1932 and was used as a command transport for Navy VIPs. Distinguished by use of blue paint on the metal hulled fuselage and wingtip floats, the single XRD-1 is shown visiting Floyd Bennett Field in New York. Powered by two Wright Whirlwind engines each of 400 horsepower, the XRD-1 grossed out at four tons and had a high speed of about 145 mph.

DOUGLAS RD-2 (H.Thorell) One of three RD-2 Dolphins for the Navy in 1933 painted up ready for use as a VIP transport. The Dolphins were normally eight place, but could be outfitted for more spacious accomodation for ranking officials. The hull was metal and the 60 foot span wing was constructed of wood. The amphibian main wheels just swung up sideways to clear the water when required. RD-2s had two Pratt and Whitney Wasps of 450 horsepower each.

DOUGLAS RD-2 (Mfr.) A company photo of March 30, 1933 shows a silver colored Navy RD-2 Dolphin and provides a good view of hull contours. The greater Wasp engine power allowed an increase of gross weight to 9400 pounds and speed increased to over 150 mph maximum. A clean amphibian design for its time, the Dolphin was a popular and useful aircraft. One went to the USMC.

DOUGLAS RD-3 (Mfr.) A little heavier and a bit slower than the RD-2, the RD-3 Dolphin was the final version procured by the Navy with six aircraft ordered and delivered in 1933-34. This RD-3 is pictured in a Douglas factory photo of October 13, 1934 prior to service squadron assignment. Some went to Navy utility squadrons and two were operated by the Marines.

DOUGLAS RD-3 (H.Thorell) One of the RD-3 Dolphins operated by the Navy utility squadrons, in this case VJ-1, in a view showing cockpit and cabin transparencies and support struts for the two Wasp powerplants along with the connecting upper airfoiled surface. The long landing gear shock strut collapsed upon gear retraction and wheels swung up around the other two struts.

DOUGLAS RD-4 (Mfr.) The first of ten RD-4 Dolphins numbered 130 through 139, is shown in a November 3, 1934 company photo just before delivery to the US Coast Guard. The author remembers seeing one or two of these USCG Dolphins at the Salem, Ma. air station in the late 1930s both on the ramp, taking off or landing in Salem Harbor, and flying over the coastal waters.

DOUGLAS RD-4 (H.Thorell) Another Coast Guard RD-4 Dolphin, this one pictured after the 1936 redesignation of USCG aircraft from 130 through 139 to V125 to V134. The aircraft shown is V126, originally number 131, presumably. Performance with the Wasp engines differed little from the RD-2 and RD-3 models. These Coast Guard aircraft were used for air-sea rescue pre-war and for patrol work during the early World War II days.

DOUGLAS R2D-1 (USN) A total of five R2D-1 transports like the civil DC-2 were purchased by the Navy starting in 1934 in an effort to keep up with the then-latest in air transport. These twin engined all metal low wing monoplanes were used by both Navy and Marines, the latter in Squadron VMJ-1, with some lasting into World War II. They had Wright 710 horsepower Cyclones giving the 18200 pound gross weight R2D-1 a top speed of about 200 mph. Wingspan was 85 feet and length 61 feet nine inches.

DOUGLAS R2D-1 (USN) There were 200 planes of the DC-2 type built by Douglas including spare parts. The majority went to civilian airlines (130 aircraft), 57 to the Army, eight were in parts, and only five went to the Navy and Marines. Funds for big Navy planes were scarce in the depression of the 1930s.

DOUGLAS R3D-1 (Mfr.) One of three Douglas DC-5 16 passenger transports procured by the Navy in 1939 is shown. Of these one crashed on its initial flight apparently because of aileron cross-rigging. The Navy later acquired a replacement as an R3D-3; this was the DC-5 that William Boeing bought. One R3D- was also destroyed during the Japanese attack on Pearl Harbor. Intended DC-5 production facilities were used for military DC-3 wartime production. Powered by two Wright Cyclones of 900 horsepower each, the approximately ten ton gross weight R3D-1 had a high speed of about 220 mph.

DOUGLAs R3D-2 (Mfr.) Pictured in a Douglas El Segundo Division photo of August 26, 1940 is a R3D-2 version of the civil high wing DC-5 for the Marine Corps. Assigned to Squadron VMJ-2 were four aircraft to be employed as cargo planes or for parachute troop training. Of the Navy and Marine R3D- aircraft six survived the war and were scrapped in 1946.

DOUGLAS R4D-5/C-47H (R.Stuckey) A postwar photo of the Douglas DC-3 in naval form, an R4D-5 Skytrain similar to the USAAF C-47A of the Norfolk, Va. Naval Reserves. Over 230 of this variant were obtained by the Navy during World War II and used primarily as cargo transports. Other -5s were used in a variety of missions. Powered by two Pratt and Whitney R-1830 Twin Wasps of 1200 horsepower each, the 95 foot span 29000 pound gross weight R4D-5 had a high speed of about 225 mph and cruised at 140 mph. In 1962 the R4D-5 became a C-47H.

DOUGLAS R4D-6/C-47J (F.Dean) A refurbished DC-3 type is shown painted up as a R4D-6 of the Naval Air Transport Service, started in 1941. The photo was taken in October, 1988. The R4D-6 was the naval equivalent of the Army C-47B Skytrain but with low altitude rated Pratt and Whitney Twin Wasp engines. Performance and weights were generally similar to the R4D-5s. In the 1962 mass redesignation the remaining R4D-6s became C-47J aircraft. About 150 R4D-6s were procured by the Navy.

DOUGLAS R4D-7/TC-47K (R.Besecker) About 40 R4D-7 transports, a training version of the Skytrain and equivalent to the Army's TC-47B were obtained by the Navy during wartime. Indicative of the fact they lasted a long time is this photo of a TC-47K taken in September of 1963 at NAS Lakehurst, N.J. Again equipped with Pratt and Whitney Twin Wasps driving Hamilton Standard propellers, the aircraft was in continuous service for at least 20 years.

DOUGLAS R4D-8/YC-47F (R. Besecker) Shown at NADC Patuxent River, Md. in July of 1967 is the prototype Super DC-3 from Douglas designated by the USAF as a YC-47F and by the Navy initially as an R4D-8. A completely updated DC-3 type in terms of airframe, powerplant, and equipment, this plane was not ordered by the USAF but after testing the Navy ordered almost a hundred of their earlier model R4D- types re-manufactured as R4D-8s, later C-117Ds.

DOUGLAS R4D-8T/TC-117D (J.Weathers) A Navy training TC-117D (after 1962) based at NATTC Lakehurst, N.J. makes an appearance at NAS Memphis, Tn. on November 6, 1976. A training version of the Super DC-3, the TC-117D shows the lengthened fuselage, new tail, new outer wing panels, and new nacelles with fully enclosed landing gear upon retraction. The increased fuselage length allowed space for a total of up to 33 passengers. The aircraft was probably serving as a navigational trainer. Wingspan was 90 feet and overall length 67 feet nine inches.

DOUGLAS R4D-8Z /VC-117D (Unknown) Some Super DC-3s were used as transports for VIPs like the one shown here. Angular lines readily distinguished the C-117s from C-47 types. Powered by two Wright Cyclones each of 1475 horsepower the approximately ten ton gross weight Navy Super DC-3 had a high speed of about 270 mph and cruised a few miles per hour faster than the high speed of the earlier ordinary R4D-/C-47 types.

DOUGLAS R5D-1Z/VC-54N (USN) First delivery of a four engine C-54 Skymaster was made to the USAAF on July 26, 1942 after a first flight on February 14. All civil DC-4 production was thus devoted to military use. Of the 252 C-54As built in California and Chicago 56 aircraft were assigned to the Navy as R5D-1s, one of which is shown in a Navy flight photo of March 21, 1944 in VIP transport form called an R5D-1Z. The airplane was for the Secretary of the Navy. It was unpressurized and had space for a crew of four and 30 passengers.

DOUGLAS R5D-2/C-54P (R.Besecker) The Navy R5D-2 version of the Douglas Skymaster is shown in August of 1962 at Floyd Bennett Field on Long Island, NY. One of 30 R5D-2 equivalents of the Army C-54B, the aircraft was similar to the -1 except for re-arrangement of fuel tankage getting all fuel out of the fuselage and integrally into the wing for a total capacity of 3740 gallons. Some Skymasters were built at Santa Monica and some in a new factory at Park ridge, Il. near Chicago. Powered with four 1350 horsepower Pratt and Whitney R-2000 engines, the Skymaster had a gross weight of about 65000 pounds and a high speed of about 280 mph. In 1962 the aircraft were redesignated to C-54P.

DOUGLAS R5D-3/C-54Q (USN) A Navy flight view of the Douglas R5D-3 on December 16, 1946 is shown. This was the Navy version of the USAAF C-54D, 86 of which were constructed, all in the Chicago plant. One R5D-3 was used in Antarctic exploration in 1956. Most of the 200 or so R5D-s in Navy and Marine service during World War II operated in the Naval Air Transport Service and were utilized over the vast Pacific distances in airlanes covering over 75000 miles.

DOUGLAS R5D-4R/C-54R (Unknown) A Marine R5D-4R is shown postwar, the type originally a Navy version of the USAAF C-54E. Of a total of 125 Santa Monica built C-54Es 20 went to the Navy as R5D-4s. This version was designed for rapid conversion of payloads. It could carry 32500 pounds of cargo, 50 troops in troop-type seats, or 44 passengers in airline seats. Cabin fuel tanks were replaced by rubber fuel cells in the wing center section.

DOUGLAS R6D-1Z/VC-118B (W.Larkins) A naval version of the civil Douglas DC-6, the R6D-1Z was ordered to the extent of four aircraft and these were delivered in 1953. In 1962 they were redesignated as VC-118Bs. These planes had VIP interiors for high level officials; note the four star placard at the cockpit location in this July 20, 1963 photo of a highly polished VC-118B of Fleet Tactical Squadron VR-1. Powered by four Pratt and Whitney R-2800 engines each of 2500 horsepower driving 13.5 foot diameter Hamilton Standard propellers, the 56 ton gross weight VC-118B had a cruise speed of 307 mph and a high speed of 330 mph.

DOUGLAS R6D-1Z/VC-118B (USN) A Navy flight view of a VC-118B shows the aircraft had a white upper fuselage and polished metal fuselage below. The four aircraft were assigned to two Pacific squadrons, VR-1 and VR-21. The R6D-1 Liftmasters replaced the R5D- Skymaster aircraft. A total of 61 R6D-1 aircraft were received by the Navy starting in 1952 after a 1950 order. Many were transferred to the USAF in the late 1950s. The standard R6D-1 had a wingspan of 117 feet six inches and a length of 105 feet.

DOUGLAS R6D-1Z/VC-118B (USN) Another Navy flight photo of a VIP R6D-1Z illustrates the clean lines of the DC-6 type, sometimes called the most efficient piston engined airliner ever built. The R6D-1 was the equivalent of the USAF C-118A. The first entered service with Air Transport Squadron Five of the Fleet Logistics Air Wing, Pacific. The aircraft could carry up to 74 passengers or 60 stretchers, or 27000 pounds of cargo. Range was over 1500 miles and service ceiling 25000 feet.

BELLANCA XRE-1 (H.Thorell) The first of three civil Bellanca five to six place CH-300 or CH-400 types purchased by the Navy in 1932 is shown after the balanced rudder had been revised with slightly greater area. As noted on the fuselage, the aircraft, based at NAS Anacostia, was used for radio experimentation, and aerial supports can be seen over the wing. Powered by a 450 horse-power Pratt and Whitney Wasp engine the 4600 pound gross weight XRE-1 had a high speed of about 145 mph. A total of 32 CH-400s were built but only three went to the Naval services.

BELLANCA XRE-2 (H.Thorell) Typical Bellanca wing and broad supporting lift struts are shown in this photo of the second Bellanca light transport for the Navy, the XRE-2 version of the CH-400. The Anacostia lettering can barely be seen behind the rear lift strut. These lift struts contributed about 47 square feet to the total lifting area. The XRE-2 had a 500 horsepower Pratt and Whitney R-1340 Wasp engine , a wingspan of 46 feet four inches, and a gross weight of 4700 pounds. High speed was about 160 mph.

BELLANCA XRE-3 (H.Thorell) The third aircraft was a CH-300 Pacemaker of 1933 designated first RE-3 and later XRE-3, and used by the US Marines. The insignia on the fuselage shows the aircraft was in Squadron VMJ-1, part of Aircraft One, Fleet Marine Force, based at Quantico, Va. Again powered by a Pratt and Whitney Wasp of 450 horsepower, the 4700 pound gross weight aircraft had a high speed of over 150 mph. The Marines used it as an ambulance for a time.

KINNER XRK-1 (H.Thorell) In 1936 the Navy acquired three Kinner Envoys, a popular civil four place light transport with a designation XRK-1. These were used as VIP transport aircraft with the paint job shown. Distinguished by a wire braced low monoplane wing, panted landing gear, and reverse-sloped windscreen, the Envoys were powered by Kinner radial engines of 340 horsepower giving the 4250 pound gross weight XRK-1 a high speed of 165 mph. Wingspan was 39 feet eight inches.

MARTIN RM-1Z/VC-3A (R.Besecker) In the early 1950s the US Coast Guard felt it required the then-latest in transportation for high level officials and obtained two Martin 404 twin engine civil transports as RM-1Zs, the Z standing for VIP use. They lasted long enough in service to be redesignated VC-3As in the new designation system of 1962. Using two Pratt and Whitney R-2800 Double Wasp engines of 2100 horsepower each the RM-1Z weighed about 45000 pounds loaded and had a high speed of just over 300 mph.

LOCKHEED XRO-1 (USN) The Navy procured a single civil Lockheed Altair DL-2A in late 1931 as shown in this Navy photo of October 8, 1931. Then on the forefront of modern design with low wing, retractable landing gear, fully cowled engine, and with closed cockpit and cabin, the Altair was to be used by Assistant Secretary of the Navy David Ingalls. Powered with a 575 horsepower Wright R-1820 Cyclone engine, the 5200 pound XRO-1 had a wingspan of 42 feet ten inches and a maximum speed of about 210 mph.

LOCKHEED XR2O-1 (USN) The Navy was never adverse to sampling the latest flying equipment and received in early 1936 a single VIP transport version of the popular civil Lockheed 10A Electra as shown. The plane was used by the Secretary of the Navy. Long lasting, it survived World War II. Using two Pratt and Whitney Wasp Junior engines of 400 horsepower each, the Electra had a high speed of just over 200 mph, weighed slightly over five tons gross, and had a wingspan of 55 feet.

Lockheed XR3O-1 (H.Thorell) In early 1936 the Secretary of the Treasury, as head of the US Coast Guard, obtained use of the Lockheed Electra shown running up in the photo, this a civil model 10-B. Powered with Wright R-975 Whirlwind engines of 450 horsepower each driving two blade Hamilton propellers, the aircraft had characteristics generally similar to the Navy XR2O-1. The XR3O-1 replaced the USCG Northrop RT-1 as a VIP aircraft.

LOCKHEED XR4O-1 (H.Thorell) A single example of the Lockheed 14-H2 Super Electra civil transport was purchased by the Navy in 1939. Based at NAS Anacostia convenient for high level Washington Navy officials, the XR4O-1 provided the latest in fast transport. Using two Pratt and Whitney Hornet engines each of 850 horsepower and with a span of 65 feet six inches, the 11 ton gross weight XR4O-1 had a high speed of 230 mph.

LOCKHEED R5O-1 (USN) Keeping up with the latest in Lockheed transports the Navy in 1941 obtained three Model 18-40 Lodestars as VIP airplanes for their highest officials. Designated R5O-1s, one of these (number 4250) is shown in a Navy photo of January 4, 1941. Another Lodestar was obtained by the Coast Guard. Using two Wright GR-1820 Cyclones of 1000 horsepower each, the 18500 pound gross weight Lodestars had a maximum speed of around 250 mph.

LOCKHEED R5O-2 (USN) The single R5O-2 Lodestar, equivalent to the USAAF C-59 and an adapted Model 18-07 civil type, procured by the Navy is shown in a pre-war photo, a predecessor of many more Lodestars to follow, and the initial aircraft of the series powered by Pratt and Whitney R-1690 Hornet engines of 1100 horsepower. This model had a crew of four and could carry 14 passengers. Characteristics were similar to earlier Lodestar models.

LOCKHEED R5O-5 (R.Stuckey) One of about 40 R5O-5 Lodestars used as executive airplanes with 12 passenger accomodations is shown in a beautiful photo. The R5O-5 was equivalent to the USAAF C-60 and like that aircraft was adapted from the civil model 18-56 Lodestar with two Wright R-1820 Cyclones each of 1200 horsepower. This model had a high speed of 266 mph and cruised at 200 mph. Normal loaded weight was 18500 pounds and maximum overload was 21500 pounds.

LOCKHEED XR6O-1 (P.Bowers) The very large size of the Lockheed XR60-1 (later XR6V-1 when the company's designation letter was changed from O to V) is shown in an early photo of the Constitution. A planned successor to the company's Constellation, the XR6O-1 got only as far as two prototype aircraft with a first flight in November of 1946. With a large double decker pressurized fuselage the Constellation could carry up to 168 passengers.

LOCKHEED XR6V-1 (P.Bowers) Shown at the end of its career in the mid-1950s, the XR6V-1 Constitution displays powerplant and landing gear details. Large cargo doors were located in the lower fuselage sections forward and aft of the wing. Powered by four of the largest piston engines available, the Pratt and Whitney R-4360 Wasp Majors each of 3500 horsepower, the two Constitutions had a wingspan of 189 feet, a gross weight of 92 tons, and a high speed of about 300 mph.

LOCKHEED R7V-1/C-121J (R.Besecker) The Navy version of the famed Lockheed 1049 civil Super Constellation aircraft, the R7V-1, 50 of which were procured for Military Air Transport Service personnel and cargo transport work, was the equivalent of a USAF C-121 and when many were transferred to that service became a C-121G. In 1962 the Navy R7V-1s were C-121Js in the new designation system. The big "Connie" could carry up to 72 passengers and was powered by four Wright R-3350 Turbo Compound engines each of 3250 horsepower giving it a high speed of about 365 mph. Wingspan was 123 feet; gross weight 72 tons.

LOCKHEED R7V-1P/C-121J (R.Besecker) A single R7V-1 Super Constellation modified as a special photo plane for use in a 1962 Antarctic expedition is pictured. The beautifully contoured lines of the Super Constellation are well shown. The Wright R-3350 Turbo Compound engines used their three power recovery turbines driven by engine exhaust to feed mechanical power back into the engine and increase its efficiency.

LOCKHEED R7V-2 (USN) Two Navy Super Constellations were acquired to act as test beds for turboprop power as R7V-2s, one of which is shown in an overhead view. The engines were Pratt and Whitney T34 turboprops each of approximately 6500 shaft horsepower each driving three bladed high activity factor Hamilton Standard propellers. The only production installation of the T34 turboprop engine was in the large USAF C-133 transport.

LOCKHEED R7V-2 (USN) Flight view of the Navy R7V-2 turboprop Super Connie version shows very clean lines. Though high performance could be gained with maximum speed well over 400 mph, problems with the powerplants, including high fuel consumption, persuaded the Navy (and the USAF with their two equivalent YC-121F models) to forgo any production of a turboprop Constellation.

LOCKHEED HC-130B (UNK.) The Lockheed Hercules has a history going back to a specification issued by the USAF in 1951. First production contract was for the USAF C-130A in September of 1952, and a total of 461 C-130As and C-130Bs were made. The US Coast Guard acquired a dozen HC-130B aircraft (originally SC-130B) as air-sea rescue types. The HC-130B shown was based at the USCG station at Elizabeth City, NC. Power came from four 4050 horsepower Allison T56 turboprops. Span was 132 feet seven inches, gross weight about 68 tons, and high speed about 360 mph.

LOCKHEED KC-130F (UNK.) A version of the Hercules used by the US Marines is shown as a KC-130F (previously a GV-1). Forty-six of this variant were procured for the USMC with deliveries complete in November of 1962. It was equipped as an assault transport and carried easily removeable probe and drogue aerial refueling equipment for two jet aircraft at a time. First flight was on January 22, 1960. As a tanker it had a capacity of 3600 gallons of fuel in the cargo compartment. It was powered by four Allison T56A-7 turboprop engines of 4050 shaft horsepower. Crew was five to seven ; normal refueling speed was about 355 mph.

LOCKHEED LC-130F (R.Besecker) One of four Hercules acquired by the Navy in 1960 initially designated as a C-130BL and later redesignated LC-130F. Taken on August 23, 1969 at NAS Quonset Point, RI. the photo shows the Hercules equipped with skis as used in Antarctic service. The uses of the Hercules seem almost endless.

LOCKHEED RC-130F (R. Besecker) A Marine RC-130F was a version used for reconnaissance missions. Originally GV-1s, the designation for this Navy/Marine Hercules was changed to C-130F in the 1962 redesignation system. These C-130 types carried a normal crew of four on the flight deck with an extra man aft as required by the mission. Propellers driven by the Allison T56 engines were usually Hamilton Standard 54H60 types. Eight JATO units each of 1000 pounds thrust could also be carried.

LOCKHEED C-130G (R.Esposito) The Navy also procured a few Hercules designated as C-130G. These were utilized for logistic support of the Navy Polaris submarine-launched ballistic missile program and also in general support of Navy nuclear submarine logistics.

STINSON XRQ-1 (P.Bowers) One of two Stinson civil Model SR-5 Reliants procured by the US Coast Guard in 1935 and given the plane number of 361. A four place high wing monoplane, the aircraft was given the designation of XRQ-1 (or RQ-1) and was used for testing of aircraft radio gear. Normal fuel capacity was 50 gallons, 25 in each wing, giving a cruising range of about 430 miles. The second of the two aircraft was designated an XR3Q-1 by the Navy.

STINSON RQ-1 (H.Thorell) A photo shows the same Stinson Reliant of the Coast Guard numbered as V149 in the later system. The USCG badge is displayed on the side of the forward fuselage. Powered by a 225 horsepower Lycoming R-680 radial engine, the Reliant had a high speed of 132 mph and cruised at 120 mph. Wingspan was 41 feet and length 27 feet two inches. Normal gross weight was 3275 pounds. Provision was made for a pilot and three passengers along with 65 pounds of baggage.

FAIRCHILD R4Q-1 (R.Stuckey) The Navy/Marine counterpart to the twin engined Air Force C-119, the twin boom rear loading R4Q-1 aircraft were in fact C-119B and C types. Initial Navy acceptances were in December of 1949. A few Marine R4Q-1s of Squadron VMR-252 were used in Korea in 1950, and later VMR-253 used the aircraft. Powered initially by Pratt and Whitney Wasp Major engines of 3250 horsepower (later uprated to 3500 horsepower), the gross weight was 36 tons and a normal crew of five was carried along with 42 to 64 troops or 35 litter patients. An alternate cargo load was 15 tons. High speed was about 260 mph. The R4Q-1 served until 1959.

FAIRCHILD R4Q-2/C-119F (P.Bowers) An updated version of the Fairchild Packet of which 58 were delivered between February and May of 1953, the R4Q-2 was equipped with two Wright R-3350 Turbo Compound engines with power recovery turbines delivering 3500 horsepower each. The new powerplant installation increased range by 20%. The new -2s flew in Marine transport squadrons and Navy Squadron VR-24 into the 1960s when they were sent to Marine Reserve units and served there until the early 1970s. In 1962 these aircraft were redesignated as C-119Fs.

FORD RR-1 (J.Weathers) The single Ford RR-1 all metal trimotor is shown preserved in the Naval Aviation Museum at Pensacola, Fl. in June of 1977. Ordered a little over 40 years earlier by the Navy, it was originally known as an XJR-1 and later redesignated. Delivered in late 1927 it served as a transport until 1930. Corresponding to a civil ten place Model 4-AT, it was powered by three Wright J-3 Whirlwind engines each of 200 horsepower and had a wingspan of 68 feet ten inches.

FORD RR-3 (USN) A Navy photo shows one of three RR-3 Ford Trimotors originally designated JR-3s of 1930. Corresponding to the civil 17 place Model 5-AT-C Trimotor transports, one airplane went to the Navy as a staff transport and the other two went to the Marines, one of which is shown. They were larger more powerful aircraft than the RR-1 and used three Pratt and Whitney Wasp engines giving the 13500 pound gross weight aircraft a high speed of 135 mph. Wingspan was 77 feet 10 inches and length 50 feet three inches.

FORD RR-4 (H.Thorell) Another Marine Corps Ford Trimotor, this the single RR-4, was again a civil 5-AT-C procured by the Navy and in 1935 given to the Marine Corps who added to the ring cowls over three 450 horsepower Wasp engines by equipping it with a snappy pair of wheel pants. High speed of the 13500 pound gross weight aircraft was about 150 mph.

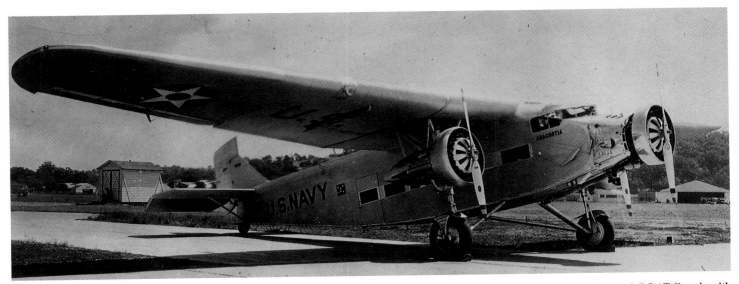

FORD RR-5 (H.Thorell) One RR-5 each went to the Navy and Marine Corps in 1935; the Navy aircraft is shown. This was a navalized civil 5-AT-C again with three Pratt and Whitney Wasp engines of 450 horsepower each giving the seven ton aircraft a high speed similar to the RR-4. The Marine RR-5 served with Squadron VJ-6M.

SIKORSKY RS-1 (USN) One of three Navy versions of the civil Sikorsky S-41 monoplane ordered in 1930, the twin engined RS-1 amphibian was for both Navy and Marines. This Navy photo taken on November 25, 1933 shows neatly cowled 575 horsepower Pratt and Whitney Hornet engines on a plane operated by Marine squadron VO-9M. The S-41/RS-1 was a larger brother to the S-38/RS-3 amphibian. Wingspan was 78 feet nine inches, gross weight 13800 pounds, and maximum speed about 133 mph with cruise speed 115 mph.

SIKORSKY RS-3 (USN) The eight place RS-3 was a redesignation into a transport category of the four amphibians originally designated PS-1 patrol planes with two gun positions. Three more aircraft were ordered later. They were Navy versions of the civil S-38B sesquiplane amphibian type. Used by both Navy and Marine Corps, the article shown on April 23, 1931 belonged to Utility Squadron VJ-1. The RS-3 was powered by two 450 horsepower Pratt and Whitney Wasp engines , had a span of 71 feet eight inches, a gross weight of about 10300 pounds, and a maximum speed of about 125 mph.

NORTHROP RT-1 (UNK.) A single example of the Northrop civil Delta 1-D was procured by the US Coast Guard in the mid-1930s. Capable of carrying five to seven passengers the single engine all metal low wing Delta, number 382 in the USCG, was powered by a 735 horsepower Wright R-1820 Cyclone engine driving a Hamilton Standard controllable pitch propeller. The plane was used as an executive transport for the Secretary of the Treasury for about a year until it was felt a multi-engined plane would be safer. Span was 47 feet nine inches, gross weight 7350 pounds, and high speed about 220 mph.

NORTHROP RT-1 (Mfr.) Another view of the Northrop RT-1 for the Coast Guard taken on February 2, 1935 at the Douglas El Segundo plant. The insignia of the USCG is already on the forward fuselage and the typical panted fixed landing gear of Northrop is shown. A fuel capacity of 328 gallons carried in six tanks in the wing center section gave the aircraft a range of 1000 miles or more.

CONVAIR RY-3 (Mfr.) The RY-3 was a transport version of the PB4Y-2 Privateer patrol bomber just as the RY-1 and RY-2 were transport versions of the earlier PB4Y-1 Liberator patrol bomber and equivalent to the USAAF C-87. A total of 38 RY-3 models using a revised fuselage from the PB4Y-2 but the same wings, powerplants, and tail surfaces were procured. Powered by four 1350 horsepower Pratt and Whitney Twin Wasp engines, the RY-3 had a wingspan of 110 feet. Gross weight was about 30 tons and high speed about 260 mph.

CONVAIR R2Y-1 (Mfr.) At the end of World War II Convair made an effort to produce a civil four engined passenger or cargo transport based on the PB4Y-2 Privateer but with a completely new fuselage of a cross-section suitable for cabin pressurization. This was the Convair Model 39 which was procured and tested by the Navy in 1945. Wings, powerplant and tail were from the PB4Y-2. No performance or weight data has been found.

CONVAIR R3Y-1 (USN) Despite the failure of the XP5Y-1 turboprop patrol flying boat program, the Navy persisted in developing a transport version known as the R3Y-1 Tradewind. Initial R3Y-1 flight occurred in February, 1954 and five boats were built. The first of these is shown in a Navy May 4, 1956 photo at NATC Patuxent River, Md. taxiing up a ramp on beaching gear using its outboard Allison T40 powerplants.

CONVAIR R3Y-2 (P.Bowers) A further development of the Tradewind turboprop flying boat was the R3Y-2 sporting a nose loading door system to discharge cargo like a beached flying LST. This view of an R3Y-2 of Navy Squadron VR-2 emphasizes the large size of the four engined boat. Handling a boat of this size was a daunting task. The tips of the dual rotation propellers can be seen above the wing. Wingspan was about 146 feet and overall length 139 feet eight inches. The T40s were each over 5800 horsepower, but these engines did not prove operationally satisfactory.

CONVAIR R3Y-2 (Mfr.) A company photo shows one of six R3Y-2 Tradewinds in flight. As on other Navy projects the four T40 twin power section turboprops each developing 5850 horsepower and driving dual rotation propellers through a common gearbox yielded a complex powerplant that gave considerable trouble, and after an initial flight in October of 1954 the Navy gave up on the 80 ton Tradewinds in 1958 after experimenting with them as aerial refueling tankers for jets. High speed was just under 400 mph.

CONVAIR R4Y-1/C-131F (F.Dean) The Navy in 1955 started receiving a near identical version of the popular civil Convair CV-340-71 twin engined airliner. Known as the R4Y-1, it was like the USAF C-131D. One of 36 ordered and first accepted in August of 1955 is shown at the Naval Air Development Center at Warminster,Pa. on May 21, 1977. A 44 passenger aircraft with gross weight of 47000 pounds, it was powered by two 2500 horsepower Pratt and Whitney R-2800 Double Wasp engines and was called the Samaritan, though the name was seldom used.

CONVAIR R4Y-1/C-131F (G.Williams) Clean lines of the Navy R4Y-1 Samaritan are shown in this September 28,1955 photo, and gave the aircraft a high speed of 275 mph. Early aircraft like this one had no nose radar; this was added along with other improvements in the late 1950s, including structural changes allowing gross weight to climb to 52300 pounds. Several R4Y-1s were used as VIP airplanes; the last R4Y-1s were delivered in 1956. Some were used by both Navy and Marine Corps on regular military transport duties. In 1962 the R4Y-1s were redesignated as C-131Fs.

CONVAIR R4Y-1Z/C-131F (J.Weathers) The one and only R4Y-1Z was ordered by the Navy in 1954 and had originally been built as a civil Model 340-66 in December of 1953. It was converted to a 24 seat VIP airplane and was based at NAS Anacostia, Washington, DC starting in April, 1954. It was assigned to the Assistant Secretary of the Navy. It served until February, 1961 when it skidded off a wet runway at Bader Field in Atlantic City, NJ into the water and was stricken from service.

CONVAIR R4Y-1/C-131F (J.Weathers) An example of an R4Y-1 Samaritan converted for VIP use is shown, this the airplane of the two star (near cockpit) Chief of the Naval Reserves visiting new Orleans on April 29, 1976. These VIP airplanes had the fuselage upper half painted white while the half below the trim line was highly polished aluminum. Propellers were Hamilton Standard Hydromatic types.

CONVAIR R4Y-2/C-131G (J.Weathers) In 1957 the Navy gave a contract to Convair for two civil Model 440s designated as R4Y-2s. Manufactured in November, they were accepted the next month. Shown about 20 years later is one of these Samaritons on September 29, 1978, an aircraft of the 4th Marine Air Wing of the Marine Corps Reserve at NAS New Orleans, La. Weight and performance were generally similar to the R4Y-1.

CONVAIR YC-131H (USN) Pictured in June of 1979 at Washington, DC is one of three VC-131H Turbo-Liners powered with two Allison T56 turboprop engines. They were converted civil Model 580s but still carried their USAF serial numbers rather than Navy numbers. Conversion was done in 1965-66 as VIP aircraft. The plane shown was used by President Lyndon Johnson to commute to his ranch from Washington until 1969, then by the DC National Guard. It was still flying in the 1980s.

GRUMMAN TF-1/C-1A (R.Besecker) A Grumman C-1A Trader of Squadron VRC-40 is shown in August of 1965, ten years after its appearance as a TF-1 before the 1962 redesignation system took effect. The Trader was a derivative of the S-2 Tracker ASW type. A total of 87 were built and used as COD (carrier on-board delivery) aircraft with a new nine rear facing passenger fuselage. Cargo could also be carried. Span was 69 feet eight inches, and length 42 feet.

GRUMMAN TF-1/C-1A (H. Andrews) An interesting photo of a Grumman TF-1 Trader from Jacksonville, Fl. taken in 1960 aboard the carrier USS Independence shows how the wings unfolded. The TF-1 flew first on January 19, 1955. The 87 aircraft were delivered between January, 1955 and December, 1958. Powered with two Wright R-1820 Cyclones each of 1525 horsepower driving 11 foot three blade Hamilton Standard propellers, the 24600 pound gross weight TF-1 had a high speed of 258 mph at 4000 feet. Normal range was 960 miles.

GRUMMAN C-2A (R.Besecker) The current long range Navy COD airplane is the Grumman C-2A Greyhound derived from the E-2 Hawkeye using the same wing and powerplants with a new fuselage and tail, in fact the first C-2As were converted E-2s. The photo shows the rear ramp loaded cargo fuselage and the four vertical tail surfaces. Seating up to 39 troops, 28 passengers, 20 litters, or five tons of cargo, the C-2A was flown first in November of 1964. A totaL of 58 were built starting in February, 1965. Many were updated in 1985-90 with new radar and other avionics.

GRUMMAN C-2A (Mfr.) A Grumman photo taken on the day of its first flight, November 18, 1964, shows the tapered wing, bulky fuselage with rear upsweep, multiple tails, and twin Allison T56 4050 equivalent shaft horsepower (later uprated to 4910 ESHP) turboprop engines driving four 13 foot six inch diameter Aeroproducts or Hamilton propellers. Span is 80 feet seven inches, and gross weight 54500 pounds. The maximum speed is about 360 mph and normal range about 1600 miles.

GRUMMAN TC-4C (AAHS) The Grumman TC-4C Academe is a flying classroom version of the Rolls Royce Dart powered Grumman Gulfstream with an APQ-88 tracking radar in the extended nose. The aircraft is fitted out to train bombardier/navigators, and carries four trainees plus two pilot trainers, also two instructors and a flight crew of two. Nine airplanes were constructed with delivery between June, 1967 and May, 1968. Gross weight is 18 tons and high speed about 385 mph.

GRUMMAN VC-4A (J.Weathers) Shown at New Orleans in March of 1977 is the single Grumman Gulfstream of the US Coast Guard. First flown on July 29, 1962 and delivered in March, 1963, the 10 to 12 seat Gulfstream is configured as a VIP transport. Powered by two Rolls Royce Dart turboprop engines driving Rotol four blade propellers, the VC-4A has a wingspan of 78 feet four inches, a normal gross weight of 18 tons, and a high speed of just over 400 mph at 2000 feet.

DOUGLAS C-9B (J.Weathers) A Navy C-9B Skytrain II of Squadron VR-56 from Norfolk, Va. is shown visiting NAS New Orleans, La. in November of 1977. The Navy followed the USAF lead in purchasing C-9s, a military version of the civil DC-9, by ordering five C-9B versions as fleet logistics support transports in April, 1972. The first two were delivered to Squadron VR-30 at Alameda, Ca. on May 8, 1973. Powered by two Pratt and Whitney JT8D turbofan engines each of 16000 pounds thrust, the C-9B has a gross weight of about 100,000 pounds and a maximum cruising speed of about 565 mph.

BELL BOEING V-22 Osprey (MFR.) The primary user of this tilt ro-tor V/STOL aircraft (*above, right*) which combines helicopter and air-plane qualities will be the US Marine Corps, though other services also have requirements. The V-22 is intended to replace the CH-46 assault helicopter now in service. The twin wingtip mounted 38 foot diameter prop-rotors can fold and the wing can rotate to allign with the fuselage so the aircraft can be stowed aboard LPH and LHA aircraft carriers. The aircraft can carry a flight crew of two and 24 troops or 12 litters. Span over the rotors is 83 feet ten inches and length is 57 feet four inches. Maximum vertical takeoff gross weight is 47500 pounds and sea level high speed is 275 knots. The two engines are Allison turboshaft types.

NAVY RACER AIRPLANES

During a short but very active period in the early 1920s the Navy, together with the Army, entered the air racing field with the stated purpose of fostering new development of high speed airplanes, engines, and equipment for use in later service types, and also to encourage a healthy spirit of competition between the two services. To what degree the first objective was really achieved could be debated, but there is no doubt about the achievement of the second.

The firm of Curtiss pretty much dominated the field of racing planes in this period with both Army and Navy using their specialized products both as landplanes and seaplanes. The premier landplane race of the period was the Pulitzer Race in the US, and for seaplanes it was the international Schneider Cup Race.

Two Curtiss landplane racers, both slick little biplanes, were manufactured for the Navy under a 1921 order for planes for the 1922 Pulitzer Race and these, the CR-1 and CR-2 (for Curtiss Racer) were flown to first and third place respectively in that race, a very respectable performance.

In the 1923 Schneider seaplane race the same CR-1 modified into a twin float seaplane won at an average speed of 177 mph.

Two more Curtiss racers were built for the 1923 Pulitzer Race, these developed and refined from the earlier CR-1 and CR-2 aircraft and labeled as R2C-1s. The two racers finished first and second in the Pulitzer in 1923 making the Navy a big winner.

It had been planned to enter the R2C-1s as R2C-2 seaplanes in the Schneider Trophy Race of 1924, but that race was cancelled.

The next year, in 1925 saw the Navy buying two more Curtiss Racers, R3C-1s as landplanes and -2s as seaplanes. These craft were more powerful and still more refined, and the idea was to enter them in the Pulitzer Race as landplanes and in the Schneider Race as seaplanes. The R3C-1s of the Navy (the Army had also purchased one) won the Pulitzer, but the two Navy R3C-2s as seaplanes could not finish the Schneider Race and an Army R3C- aircraft was the winner.

An updated R3C-2, designated as an R3C-4, was entered in the 1926 Schneider Race, but was forced out and did not finish.

This was the end of the Navy's efforts to use specialized aircraft to enter racing. Any other racing types were modifications, however major, of service aircraft types such as the XF6C-6 Curtiss Page racer, a major modification of a Curtiss Hawk airframe.

CURTISS CR-1 (Mfr.) Also known at Curtiss as the L-17-1, but usually just as Curtiss Racer 1, this aircraft (A-6080) was one of two racers ordered by the Navy in June of 1921. Flown by Bert Acosta in the 1921 Pulitzer Race with the Lamblin radiators shown in the photo, the racer won at a speed of 176.7 mph. It was later flown at 197.8 mph. With new wing skin radiators as a Navy racer it placed fourth in the 1922 Pulitzer at 188.8 mph. In 1923 the CR-1 became a CR-3 seaplane. Wingspan was 22 feet, eight inches. The Curtiss CD-12 engine #2 of 410 horsepower at 2180 RPM drove a seven foot two inch diameter Curtiss propeller. Gross weight was 2158 pounds with the Lamblin radiators. The fuel was 50% Benzol.

CURTISS CR-2 (Mfr.) A photo of September 16, 1922 shows the second Curtiss racer (also the L-17-3) with Navy Lt. H.Brow. The CR-2 was equipped with skin cooling for the 1922 Pulitzer Race as shown in the photo. Lt. Brow flew it to third place in 1922 at a speed of 193.2 mph. Powered by the Curtiss CD-12 engine #4 of 410 horsepower at 2250 RPM the CR-2 had a 40 gallon fuel capacity, a span of 22 feet eight inches, a gross weight of 2212 pounds, and a high speed of 198.8 mph. Landing speed was 70 mph.

CURTISS CR-3 (Mfr.) An August 1923 photo shows the CR-1 revised into a CR-3 twin float seaplane for the 1923 Schneider Cup Race in England. Several changes besides the floats were made including use of a new Curtiss D-12 high compression engine giving 460 horsepower at 2300 RPM driving a Reed metal propeller of eight feet nine inches in diameter. Gross weight came to 2747 pounds. High speed was 194 mph. The CR-3 won the 1923 race piloted by Lt. Dave Rittenhouse at 177.4 mph. The CR-3 later was flown at 188.07 mph to set a new seaplane world speed record.

CURTISS R2C-1 (C.Mandrake) A new Navy racer from Curtiss in 1923, the R2C-1 was developed from earlier racers, and two were ordered with new skin cooling on both wings which were each tied to the fuselage directly. Shown is A-6691 with race number 10. In October, 1923 in the Pulitzer Race the R2C-1s placed first (#9) and second (#10) at respective speeds of 243.7 mph and 241.8 mph with Lt.Al Williams the winner. The R2C-1 was powered by a Curtiss D-12A high compression engine of 500 horsepower at 2300 RPM driving a seven foot 10 inch Reed metal propeller. Wingspan was 22 feet, gross weight 2098 pounds, and high speed was rated at 247.5 mph. Landing speed was 75 mph.

CURTISS R2C-2 (Mfr.) The R2C-2, also called the L-111-1 at Curtiss, and numbered A6692 by the Navy, was the second R2C-1 on twin floats. It was to be entered in the 1924 Schneider Race which was cancelled. Thereafter it was used as a trainer for Schneider racers of the two subsequent years and was wrecked in 1926. It used a Curtss D-12A engine of 502 horsepower at 2300 RPM with a Reed propeller, weighed 2640 pounds as a seaplane, and had a high speed of 226.9 mph based on a Curtiss test of September 27, 1924. Landing speed was 80 mph.

CURTISS R3C-1/-2 In 1925 the Navy bought two (of three; the Army bought the other) R3C-1 racers, a new updated design. When raced as landplanes they were R3C-1s; as seaplanes they were R3C-2s (though the photo shows a seaplane as an R3C-1). Navy pilot Lt. George Cuddihy is shown with the racer in the picture. Intended for both Pulitzer landplane and Schneider seaplane racing, two R3C-1s won the 1925 Pulitzer. The winner's speed was 249 mph.

CURTISS R3C-2 (Mfr.) The two Navy R3C-aircraft were put on twin floats for the Schneider Cup Race (as was the single Army aircraft) and in October, 1925, after the two Navy planes had to drop out of the race Lt. J.J.Doolittle of the Army won at a speed of 245.7 mph. The aircraft is now in the National Air and Space Museum. The R3C-2 was powered by a Curtiss V-1400 engine of 600 horsepower at 2350 RPM driving a Curtiss EX-32995-112 propeller of ten feet diameter. Span was 22 feet. Seaplane gross weight came to 2738.4 pounds and high speed was as noted.

CURTISS R3C-4 (Mfr.) Labeled in the company photo of October 28, 1926 (*above*) as Navy-Curtiss Schneider Cup racer R3C-4 with Curtiss V-1550 motor, the Curtiss racer was an R3C-2 updated with new powerplant and revised floats. It did not finish the 1926 Schneider race. The Curtiss V-1550 engine developed 705 horsepower at 2600 RPM and drove a Curtiss Reed R type 7 foot eight inch diameter propeller giving the aircraft a high speed of 255 mph. Gross weight was 3223 pounds.

NAVY SCOUT AIRPLANES

Scouting was a very early mission for Navy airplanes with the first domestic types appearing shortly after World War I. Spotting for naval gunfire was one of the principal functions. In the early 1920s long range scouting became a major mission of Naval Aviation whether the aircraft was ship or shore based. By 1930 this long range mission was performed mainly by patrol (VP) squadrons, but scouting (or observation) planes, assigned to shore bases, capitol ships, and carriers continued in service with missions unchanged well into the World War II period. Postwar the aircraft designated as scouts either disappeared from fleet service or were new types specifically designed and employed as carrier-based ASW planes. The pure scouting mission of the past went away as the advent of long range radar for search brought about a change in the part played by aircraft as the scouting eyes of the fleet.

One of the earliest types designated in the scout category was the Loening LS-1 two place monoplane twin float seaplane of 1921, but there was no production. A year later in 1922 an Aeromarine experimental two place twin float seaplane was tested as the AS-1 and with changes the AS-2 but stayed as only a prototype.

In 1923 Curtiss put out a large slow long range convertable type based on a Bureau design, actually a multi-purpose plane built for shore or airplane tender-based scouting operations as the CS-1 (for Curtiss Scout-1). A couple of modified CS-2s were also built as prototypes, but Martin obtained the 1925 production order by under-bidding in the manner of those times. The SC-s were the last shore or tender-based long range scouts; patrol planes took over the function in the late 1920s.

In 1923 small scout planes for use with a large submarine were built and experimented with. The Martin MS-1, a small twin float biplane was one of these designs; others were the Cox-Klemin XS-1/XS-2 design, the Loening M-2s, and a J.V.Martin K-4 along with some foreign types. None became truly operational.

Some time later in 1931 another attempt at a submarine-based scout plane was made in the form of the monoplane Loening XSL-1 (and XSL-2 two years later) with a pusher propeller and a boat hull. Only one plane of this type was built. A much more successful airplane of 1931 was the Vought scout version of the famous biplane Corsair, the SU-1 used on board aircraft carriers. Various production SU- types followed up to a -4 model and served the Navy for several years.

Scout plane prototypes abounded in 1932 and 1933. Curtiss built a late model two place biplane Helldiver appearing in 1932 designated an XS3C-1 and a year later put out an XS2C-1 low wing monoplane patterned on the Army low wing A-12 attack airplane with a fixed landing gear and no wing folding. Neither aircraft gained a production contract. A 1931 Navy competition for a new combined capital ship or aircraft carrier service type as an amphibian with folding wings resulted in some strange configurations in 1932, the Great Lakes XSG-1, a Loening XS2L-1, and the Sikorsky XSS-2. After these designs were tested all were rejected. A final Navy scout candidate of 1932 was the XSE-1 from Bellanca, a high wing monoplane which crashed and was rebuilt with changes as an XSE-2. No production Bellancas resulted from testing.

The year 1934 saw the advent of a successful Navy scout plane, the Grumman SF-1 version of the two seat Grumman FF-1 "FiFi" fighter. The SF-1 scout went into squadron service.

The only production World War II Navy plane in the scout category was the Curtiss single seat SC-1 high performance low wing monoplane convertable for land or catapult use off major combat ships. Used late in the war the SC-1 Seahawk was the final Navy catapult scout aircraft and lasted into the late 1940s.

Starting in the 1950s carrier based ASW aircraft were designated in the scout category and in 1953 the first of the long-lived Grumman twin engined S2F- Trackers (later the S-2 in 1962 and on) appeared, and was produced in quantity as the first line aircraft of its type into the early 1970s. Many were later converted into utility planes as US-2s.

The current Navy scout aircraft is the Lockheed S-3 Viking carrier based ASW monoplane using twin turbofan engines. The S-3 first appeared in 1973, and is still in use in the 1990s. No replacement has yet appeared, and the Viking has received frequent updates.

AEROMARINE AS-2 (Mfr.) A company photo of March 10,1922 shows an Aeromarine Plane and Motor Company AS-2 scout or observation plane, one of three delivered that year as AS-1s. A twin float double bay wing seaplane with a forward fixed and rear flexible machine gun, the AS-1s became AS-2s with powerplant revisions. Powered by a 300 horsepower Hispano Suiza engine and with a wingspan of 37 feet six inches, the 3600 pound gross weight AS-2s had a high speed of about 115 mph.

CURTISS CS-1 (Mfr.) A Curtiss photo of November 26, 1923 shows a new CS-1 designed for the primary mission of scouting and alternate missions of bomber or torpedo plane. The large 56 foot six inch span singla bay biplane was designed to replace the Douglas DT- series aircraft and could be converted to a twin float seaplane. The upper wingspan was shorter than the lower to make all wing panels similar. Six aircraft were delivered in 1923-24. The engine was a water cooled Wright T-2 of 525 horsepower giving the four ton landplane version a high speed of about 100 mph. The seaplane was heavier but not much slower.

MARTIN SC-1 (Wright) A Martin three place SC-1 scout plane is seen running up. The reason the aircraft looks like the Curtiss CS-1 is because, in the custom of that day, the CS-1 design was let out on bids for production articles, and Martin under-bid Curtiss. Martin produced 35 SC-1s in 1925. The SC-1 scout was powered by a Wright T-2 engine of 525 horsepower, weighed 8300 pounds as a landplane, and had a maximum speed of about 100 mph whether in landplane or seaplane configuration.

MARTIN SC-1 (USN) Navy flight view of a Martin SC-1 scout of Torpedo Squadron VT-1 configured as a twin float seaplane shows the cockpit for the third crewman just aft of the Lewis machine gun mounted in a Scarff ring in the middle cockpit. The photo was taken on September 26, 1925. Note the biplane wings had no stagger, typical for some planes of the period. Ailerons were on both wings with a drive strut between them. The seaplane was about 700 pounds heavier than the landplane version.

CURTISS CS-2 (Mfr.) One of two prototype Curtiss CS-2 scouts shown in a company photo of January 29, 1924. It was similar to the CS-1 except for use of an uprated Wright T-3 engine installation of 585 horsepower and additional fuel tankage. The nose lines are clean because the water cooling radiator is on the other side of the powerplant. The split axle landing gear allowed carriage and launching of a torpedo from under the fuselage. The landplane version had a gross weight of 10360 pounds and a high speed of about 105 mph.

MARTIN SC-2 (USN) Martin again outbid Curtiss to build the CS-2 version as an SC-2 with the designation letters reversed to conform to the new system. One of 40 Martin SC-2 seaplane scouts, this in Navy Squadron VS-1 is shown dropping a long 1600 pound torpedo. The Martins were delivered in 1925 and used the same Wright T-3 engine as the Curtiss CS-2s. Seaplane gross weight was about 9300 pounds; speed was still around 100 mph. The SC-2 scouts were later redesignated as T2M-1 torpedo planes.

CURTISS SC-1 (Mfr.) The Curtiss SC-1 World War II Seahawk was a catapult scout plane for cruisers or battleships. The company photo at the Columbus, Oh. plant shows the landplane version as Curtiss delivered them. The single main and tip floats were contracted for separately by the Navy from the Edo Corporation. Curtiss obtained a contract for two XSC-1 prototypes in March of 1943 and for 500 production SC-1s later that year. First flight was on February 16, 1944. Armament of the single seat SC-1 consisted of two .50 caliber guns fixed in the wings just outside the propeller arc; wings folded aft into a compact package for stowage. Two store stations could carry bombs, tanks, or a radar pod.

CURTISS SC-1 (H.Levy) A photo of July 24, 1947 shows an SC-1 Seahawk used as a utility aircraft. Most were out of service a year later, either mothballed or sent to the smelter. First production delivery of the SC-1 was made to the USS Guam on October 22, 1944; the type first went into Pacific action just before war's end in June of 1945. A great amount of difficulty was encountered in SC-1 spin recovery testing and the type was restricted from deliberate spinning. A total of about 550 Seahawks were delivered.

CURTISS SC-1 (Mfr.) A Curtiss photo shows a flight of SC-1s before delivery. Wings incorporated leading edge slats. Wing guns and stores racks can be seen. Some of the landplanes were equipped with semi-panted landing gear. The engine was a 1350 horsepower turbo-supercharged Wright Cyclone and the propeller was a four bladed Curtiss Electric type of ten feet two inches diameter. Span was 41 feet; gross weight as a landplane was 7240 pounds, and high speed was about 315 mph.

CURTISS SC-1 (USN) A Navy photo shows an SC-1 as a single main and tip float seaplane used on catapults of major combat ships and superseding the OS2U-, SOC-, and SO3C- types with a plane of much increasd performance. It took about six hours to convert a Seahawk to a seaplane configuration. The SC-1 was the last shipboard seaplane scout in the Navy and was replaced by long range radar and helicopters. Gross weight increased about 700 pounds to 7935 pounds as a seaplane. It was said a Seahawk could outclimb an F6F- Hellcat to 6000 feet and out-turn an F8F- Bearcat.

CURTISS SC-2 (Mfr.) A Curtiss photo shows the revised SC-2 Seahawk model. A large contract for this type was cut back at war's end in August, 1945 and only ten SC-2s were delivered after production of two XSC-2s. A new 1425 horsepower version of the Wright R-1820 engine powered the SC-2 enclosed in a new circular cowl. Also added was a full blown bubble canopy. The picture shows a radar pod under the starboard wing and a fuel drop tank under the belly of the fuselage.

CURTISS SC-2 (Mfr.) Flight photo of the SC-2 shows the new bubble canopy, new radio mast location, and the forward extensions to the horizontal stabilizers, the latter part of an effort to improve the notoriously poor SC- spin characteristics. When the last of the SC-2s was delivered to the Navy in October of 1946 the end of the Navy-Curtiss aircraft relationship was very near.

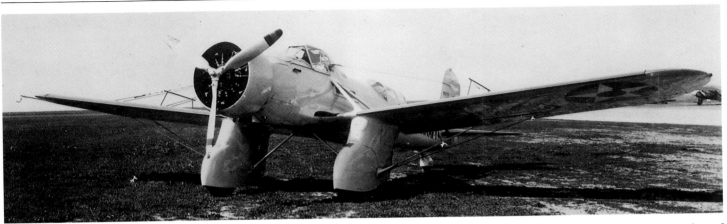

CURTISS XS2C-1 (Mfr.) An unusual Navy scout monoplane was acquired from Curtiss in December, 1932. This was the XS2C-1 shown in a company photo of June 5, 1933. A version of the Army's A-8 attack plane, but using a Wright R-1510 625 horsepower radial engine instead of an inline Conqueror, the XS2C-1 had a wingspan of 44 feet and the wing did not fold, making it unfit for carrier basing and possibly only useful to the Marines. No further aircraft were procured however. Gross weight was about 4800 pounds and high speed approximately 185 mph.

CURTISS XS3C-1 (Mfr.) A cleaned up Helldiver with a Cyclone engine, the XS3C-1 scout is shown in a photo of January 29, 1932. It had a short life when it crashed during a test flight at NAS Anacostia in late February. The single strut panted landing gear and carefully filleted "N" struts are noteable. Compared to other Cyclone powered Helldivers the open cockpits seem like a retrograde step.

CURTISS XS3C-1 (Mfr.) Handsome company flight photo of the XS3C-1 Cyclone Helldiver on the same day as the previous picture was taken, and prior to Navy testing. Curtiss data of February 27, 1932 indicate the plane was intended as a carrier scout or light bomber and was powered by a Wright R-1830E Cyclone of 620 horsepower direct driving a Hamilton Standard propeller of ten feet diameter. Wingspan was 32 feet and fuel capacity was 120 gallons. As a scout gross weight was 4941 pounds and high speed 180 mph; as a bomber with about 250 pounds of bombs the gross weight was 5098 pounds.

BELLANCA XSE-1 (Mfr.) In October, 1931 the Navy ordered a single monoplane scout from Bellanca with panted fixed landing gear and lift struts. The company photo of December 15, 1932 shows the unusual configuration. Equipped with a tail hook but with a fixed wing of 49 feet nine inches span it is clear the plane would not fit well aboard carriers. Powered with a 650 horsepower Wright R-1820 Cyclone , the XSE-1 had a gross weight of about 5530 pounds and a high speed of around 170 mph.

BELLANCA XSE-2 (H.Thorell) After an accident the XSE-1 scout was rebuilt in 1933-34 by Bellanca resulting in the XSE-2 shown. The major changes from XSE-1 were installation of a twin row Wright R-1510 engine of 650 horsepower providing a slimmer nose contour and revised tail surfaces with more area. Again the Navy saw the light after tests and rejected the XSE-2 scout; it was not carrier-compatible in any case.

BELLANCE XSE-2 (UNK.) Another view of the XSE-2 shows a change in contour of the landing gear "pants" and the installation of a rear flexible .30 caliber machine gun. The tail hook for carrier arrested landings is beneath the rudder. The long greenhouse could easily house a crew of three. The lower "lift strut" system was typical of Bellanca design.

GRUMMAN XSF-1 (Mfr.) A single two seat scout prototype similar to the FF-1 fighter is pictured. The XSF-1 aircraft was novel for use of a retractable landing gear and enclosed cockpits at the time. First flight was on August 30, 1932 and the plane was delivered to the Navy the same day. With respect to the FF-1 fighter version one of the fixed machine guns and its ammunition was removed to provide space for more fuel capacity. It initially used a 600 horsepower R-1690 Pratt and Whitney Hornet engine but this installation was changed to a Wright R-1820 Cyclone of 750 horsepower giving the 4376 pound gross weight XSF-1 a high speed of just over 200 mph.

GRUMMAN SF-1 (G.Williams) Two fine views (*above, right*) of the production SF-1 scout, 33 of which were purchased by the Navy based on successful tests of the XSF-1 prototype. The first production aircraft had its initial flight February 15, 1934 and production was completed by July 12. The SF-1 had one fixed and one flexible .30 caliber gun and could carry two 100 pound bombs. Powered by a 750 horsepower Wright R-1820 Cyclone engine driving a nine foot six inch diameter Hamilton propeller, the 5070 pound gross weight SF-1 had a high speed of 206 mph.

GRUMMAN XSF-2 (R.Besecker) This airplane was the last production SF-1 scout modified to have a Pratt and Whitney R-1535 twin row radial engine of 650 horsepower in a long chord cowling driving a Smith two blade propeller of eight foot six inch diameter. Besides having armament as the SF-1 the XSF-2 had provision for installation of a 500 pound bomb cradle between the landing gear legs. Note the aileron drive strut canted to drive upper wing ailerons only.

GRUMMAN XSF-2 (H.Thorell) Another view of the XSF-2 with the close cowled R-1535 engine including bumps for rocker box covers on the front row of cylinders. The plane was based at NAS Anacostia. Md. as noted on the fuselage. First flown on November 26, 1934 the aircraft was delivered to the Navy December 3, 1934. Wingspan was 34 feet six inches, gross weight 4790 pounds, and high speed 215 mph.

GRUMMAN S2F-1/S-2A (R.Besecker) An S-2A (was S2F-1 before 1962) Tracker of the Naval Reserves at Minneapolis, Mn. is shown, the initial version of the first single Navy type designed as both hunter and killer in the ASW role. After two XS2F-1 prototypes were tested with first flight in December, 1952 a total of 755 Trackers were to follow between October, 1953 and March of 1960.

GRUMMAN S2F-1/S-2A (Mfr.) A company flight view of the S2F-1. The four place aircraft had two pilots and two electronics operators. A radome above the cockpit, a searchlight under the starboard wing, store stations under the port wing, and retracted MAD (magnetic anomoly detection) gear in the tail are shown. A retractable radome was housed in the fuselage belly behind the torpedo/weapons bay. Sonobuoys were inside the aft section of the engine nacelles. The S2F-1/S-2A was the standard carrier based ASW airplane of the 1950s. Wingspan was 69 feet eight inches. but the wings folded to just over 27 feet.

GRUMMAN S2F-1/S-2A (USN) Two S2F-1 Trackers of the Oakland Reserves fly in over the California coast. the bulge of the retracted belly radome can be seen under the fuselage. Powered with two Wright R-1820 Cyclones of 1525 horsepower each driving a three blade Hamilton Standard propeller of 11 feet diameter, the 24500 pound gross weight S2F-1 had a high speed of about 270 mph.

GRUMMAN S2F-1T/TS-2A (R.Besecker) Some of the S-2 aircraft were revised into trainer versions as time went on; one of these is shown in the photo. The searchlight on the wing has been retained but radomes have been removed, so the aircraft was not used for ASW operational training.

GRUMMAN US-2A (F.Dean) A photo taken May 21, 1977 at NAS Willow Grove, Pa. shows a Reserves S-2 used as a utility hack years after retirement from front line service with ASW tactical equipment removed. The interesting method of wing folding is shown. Both wings folded up over the fuselage, but the fold joints are angled so that the panels miss each other in the folding process.

GRUMMAN US-2B (R.Besecker) The S-2B was an improved S-2A with updated ASW electronics (Julie/Jezebel systems). An October, 1966 photo shows an S-2B retired into utility use as a US-2B. The wing fold joint is well shown with means of routing necessary services through the joint indicated. In addition the tail bumper shows and the belly radome is still in the retracted position.

GRUMMAN US-2C (R.Besecker) A September, 1955 photo shows a US-2C utility version in Navy Squadron VC-5. The aircraft was originally an S2F-2 Tracker in active ASW service (S-2C in 1962). A total of 60 S2F-2s were built with the first flying in July of 1954 and production extending into January of 1955. This version had an enlarged bomb bay and a wider tail. Bomb bay doors are open in the photo. Gross weight of the S2F-2 was about 25500 pounds and high speed about 260 mph.

GRUMMAN US-2C (S.Orr) A Navy flight view shows a utility modification of an original S2F-2 aircraft serving in Utility Squadron VU-3 probably being used as a transport. All original ASW tactical equipment has been removed but the housing for the retractable belly radar installation can be seen.

GRUMMAN S2F-3/S-2D (R.Besecker) The photo of the S-2D of Scouting Squadron VS-30 was taken in September of 1965. The S-2D was an S2F-3 between its first flight of May, 1959 and 1962 when the new designation system was started. A total of 119 of this type were built between July, 1959 and December, 1967. Electronic equipment was updated and more added, and provision was made for carriage of a nuclear depth charge. Fin height was increased, wingtips were rounded, and powerplant nacelles revised. High speed was about 220 mph and gross weight 29150 pounds.

GRUMMAN S2F-3/S-2D (USN) Taxi photo of an S-2D illustrates the fuselage bulges made to increase the size of the belly weapons bay on this and subsequent versions to accomodate two torpedoes instead of one and a nuclear depth charge. There are three store stations under each wing.

GRUMMAN US-2D (R.Besecker) Like earlier versions of the Tracker the S-2D was often revised into a utility hack for general transport duties, and an aircraft out of front line service and based at NAS Norfolk, Va. is so used here. The author recalls an S-2D used to transport several Navy personnel to his plant from Washington, DC. for a meeting.

GRUMMAN S2F-3S/S-2E (S.Orr) Another Navy flight photo shows an S-2E Tracker of which 241 were produced. An initial flight in September of 1960 was followed by production between July 11, 1962 and December, 1967. The S-2E was an S-2D with revised electronics such as a new automated tactical navigation system. Fourteen of these were exported to Australia.

GRUMMAN S-2E (R.Besecker) An airshow photo of an S-2E Tracker of Scouting Squadron VS-24 is particularly interesting since it shows an operational aircraft with the belly radome in the lowered search position, the tail hook lowered (this combination would not happen at the same time) and the MAD gear extended from the tail in a position as far as possible from the metal engines mass.

GRUMMAN S-2F (R.Besecker) An S-2F Tracker sits with folded wings. This designation applied to some early S-2B models with the electronics suite updated. All S-2F aircraft carried the Julie/Jezebel systems. In addition the S-2G model was an S-2E with new electronics. In the early 1970s all the S-2s were becoming outdated and a new carrier-based ASW aircraft system was needed, resulting in the S-3A from Lockheed.

GREAT LAKES XSG-1 (H.Thorell) An unusual configuration was shown for an intended carrier or catapult scout amphibian plane by the Great Lakes XSG-1 appearing in late 1932. A two place biplane with retractable wheels and the rear seat man below the pilot, the single XSG-1 prototype did not measure up to Navy requirements. Wingspan was 35 feet. Powered by a Pratt and Whitney R-985 Wasp Junior engine of 400 horsepower, the 4220 pound gross weight amphibian had a high speed of about 130 mph.

LOENING LS-1 (USN) Shown in an April, 1921 Navy photo is a two place LS-1 twin float seaplane version of Loening's M-8-O monoplane observation plane, one of many variants of a Loening basic design. The LS-1 was distinguished by a strange float design with flat inboard sides as if a single main float had been sliced in half longitudinally. Only one aircraft was built. Powered by a 300 horsepower Hispano Suiza engine, the small 2500 pound gross weight seaplane had a high speed of about 120 mph. Note the Scarff ring mount for a rear flexible gun.

LOENING XSL-1 (NACA) Another Loening monoplane scout, this one intended to fold and fit inside the hangar of the large submarine S-1, the single place XSL-1 is shown in an NACA photo during testing in early 1931. Powered by a 110 horsepower Warner Scarab radial engine mounted well above the boat hull on struts and driving a pusher propeller, the three quarter ton gross weight XSL-1 could fly at speeds up to about 100 mph.

LOENING XSL-2 (USN) A Navy photo of May 24, 1933 at Anacostia, Md. shows another attempt to make the Loening submarine scout flying boat successful by substituting a six cylinder Menasco engine to drive the pusher propeller. How the aircraft folded to fit into the submarine hangar is not clear. The 31 foot span XSL-2 was a little heavier and faster, but the effort was not successful and no more were built.

LOENING XS2L-1 (USN) Differing from the earlier Loening scout prototype in being a tractor biplane amphibian was the XS2L-1 of early 1933 using Grumman-style wheel retraction into the hull. An enclosed cabin for pilot and gunner was featured with the pilot looking directly through the propeller disc. The XS2L-1 was another single prototype ; it was powered by a 400 horsepower Pratt and Whitney Wasp Junior engine, had a wingspan of 34 feet six inches, a gross weight of about two tons, and a maximum speed of about 130 mph.

MARTIN MS-1 (Mfr.) An early attempt at a submarine-based Navy scout plane was the Martin MS-1 of 1923, a tiny twin float biplane with a wingspan and a length each of only 18 feet! The metal floats had plenty of support struts! How long it took to detach floats and fold wings for submarine stowage is not known. Powered with a Lawrence three cylinder engine of 60 horsepower, the MS-1 grossed out at only 1000 pounds and could attain a high speed of 100 mph. Six were built by Martin.

SIKORSKY XSS-2 (USN) Another light scout amphibian prototype of 1933 was a Sikorsky product, the XSS-2 shown in a May 24, 1933 photo. Of unusual configuration with a gulled monoplane wing on a boat hull and tip floats, the two place XSS-2 shows a carrier arresting hook behind its tail wheel and aft-located cockpits yielding what must have been poor forward visibility.

SIKORSKY XSS-2 (USN) A slightly later Navy photo of June 15, 1933 shows the XSS-2 scout prototype in flight with wheels neatly retracted and a cleanly cowled engine high on struts. Powered with a Pratt and Whitney R-1340 Wasp engine of 550 horsepower driving a tractor propeller, the 4500 pound gross weight Sikorsky had a top speed of about 150 mph. The aircraft remained a single prototype while the Navy opted for more conventional scouting planes.

VOUGHT SU-1 (H.Thorell) The scout versions of the Corsair in the SU- series were more powerful and normally a little heavier than the O2U- series and were initially called O3U-s. They were carrier based and came out later than the O2U- types. The photo shows an SU-1 sharped up with wheel pants and carefully filleted wing strut terminations along with a rear seat headrest and a blue paint job over the fuselage for use as a command aircraft.

VOUGHT SU-1 (H.Thorell) Another of the 29 SU-1 Corsairs ordered in early 1931 and delivered from late that year to early 1932. These aircraft were originally procured as O3U-2 observation planes. Again the SU-1 running up in the photo has been gussied up with wheel pants. In this case the single fixed gun is shown in the upper wing and the rear cockpit is set up for a flexible gun installation. Note the bulged fairing on the fuselage for flotation gear and the tail hook fitted. The standard SU-1 had a 600 horsepower cowled Pratt and Whitney R-1690 Hornet engine. Span was 36 feet, gross weight about 4500 pounds, and high speed about 170 mph. Marine squadrons used the SU-1.

VOUGHT SU-2 (H.Thorell) The SU-2 was originally designated an O3U-4 before changeover to a scout designator. The Corsair shown is a rear admiral's command aircraft as indicated by the two star placard alongside the rear cockpit. The admiral could land on his carriers, but there were no guns on the aircraft to shoot! Note the tail wheel, standard on the SU- types. A total of 44 SU-2 airplanes were produced; all were landplanes. Except for a different Hornet engine version, specifications and performance were essentially the same as for the SU-1.

VOUGHT SU-2 (Mfr.) A manufacturer's photo taken on July 28, 1932 shows a command plane for admiral's use ready for delivery from the factory. This is the same plane pictured in the previous photo. The outside metal surfaces of the fuel tank can be seen on the fuselage side inboard of the flotation gear housing, typical of Vought design.

VOUGHT SU-3 (R.Besecker) The SU-3 Corsair arrived in 1932 and 20 of this model were purchased to be used for photgraphy purposes. Like the SU-2 the SU-3 was originally designated an O3U-4. This fine Corsair photo shows the location of the single fixed .30 caliber gun in the upper wing; the rear flexible gun is not mounted. Also in evidence is the tail wheel and carrier hook, the outside panels of the fuel tank contoured to basic fuselage lines, and the protruding bulge of the flotation package.

VOUGHT SU-3 (H.Thorell) After serving aboard carriers and with the Marines in operational squadrons some SU-3 Corsairs were employed as VIP transports and utility planes. This highly polished sample based at NAS Anacostia and running up its Hornet engine is distinguished by having a hatch to cover the front cockpit and form a partial enclosure over the rear. Undoubtedly a command transport it has no carrier hook fitted.

VOUGHT SU-3 (H.Thorell) An interesting photo of an SU-3 Corsair used by Navy Bombing Squadron VB-5, probably as a utility plane. Also interesting is the use of a rudder with a shape like an SU-4 but marked as an SU-3. As seen in the photo the location is Floyd Bennett Field in New York. The SU-3 had a wingspan of 36 feet, a normal gross weight of just over 4500 pounds, and a high speed of approximately 170 mph.

VOUGHT XSU-4 (Mfr.) A Vought photo of September 18, 1933 shows the initial SU-4 Corsair, sometimes called an XSU-4, though this is not marked on the curved rudder of the aircraft. In pristine factory-new condition, this scout has not yet been given a squadron assignment. It is believed the aircraft ended up at NAS Anacostia, Md. Like earlier SU-scouts the SU-4 used a 600 horsepower Hornet engine; it was a little heavier and slower than an SU-3.

VOUGHT SU-4 (H.Thorell) One of 20 SU-4s produced serving with a bombing squadron, VB-5, is pictured in a wintry scene. SU-4s were the final scouting-type Corsairs; the final Corsairs had yet to come as O3U-6 observation planes. The time of the open cockpit biplane was near an end. Note the absence of a tail hook. The SU-4 was distinguished by a slight forward extension of the vertical fin.

LOCKHEED S-3A (R.Dean) The most modern current carrier-based ASW airplane is the Lockheed S-3A Viking shown in a late 1970s photo at NAS Willow Grove, Pa. This is the four place replacement for the piston engined Grumman S-2 and is powered by two turbofan engines. Lockheed was given the S-3A contract in mid-1969 and a Viking first flight took place in January of 1972 as one of eight service test aircraft. First production orders were given later that year.

LOCKHEED S-3A (USN) An S-3A Viking of Squadron VS-38 is shown on the deck of the carrier USS Enterprise, CVN-65, in the Pacific on April 13, 1978 during RIMPAC '78 Exercise along with an EA-6B Intruder of VAQ-134 and Tomcats of VF-1. The S-3A configuration is well shown with the General Electric TF34 turbofan engines each of 9300 pounds thrust inboard and stores pylons outboard, the latter in this case fitted with a drop tank. Wing leading edge slats are open. The wingtip items are ECM antennas. The cockpit of an S-3A is, the author can attest, like something right out of Star Wars!

LOCKHEED S-3A (USN) An S-3A Viking comes in for a carrier landing during trials with arresting hook deployed to catch a wire and with flaps down. Stall speed was normally about 70 mph. The Viking first went to Navy squadrons in 1974. It has a wingspan of 68 feet eight inches which folds to just under 30 feet. Gross weight is normally 43500 pounds but can go up to 26 tons in overload condition. High speed is about 510 mph.

LOCKHEED S-3A (USN) A Navy flight photo of September 23, 1974, a few months after the Viking first entered service, shows an S-3A of VS-41, the first unit to use the aircraft. Flying off the coast of San Diego, this airplane was based at NAS North Island. The diagonal line on the vertical fin shows the fold line of that surface. The MAD boom is fully retracted into the tail. A belly weapons bay can carry four torpedoes, bombs, or mines. A large number of sonobuoys can also be carried along with various weapons on wing store stations.

NAVY SCOUT BOMBER AIRPLANES

The Navy scout bomber (VSB) aircraft category started in the mid-1930s when the idea of dual role aircraft became popular. There were, including experimental prototypes and limited production models, a total of a dozen scout bombers. After World War II the scout bomber category died out and, together with the torpedo bomber type, was replaced briefly by the VBT bomber torpedo category and later simply the attack designation.

The first Navy airplane put in the scout bomber category was the Curtiss XSBC-1 of 1934, a two seat high wing strut braced monoplane with retractable landing gear and a modification and redesignation of the earlier XF12C-1 two seat fighter prototype. A crash ended the career of the XSBC-1.

A year later in 1935 Vought brought out their SBU-1 scout bomber, another conversion of an earlier experimental two seat fighter prototype, this one with fixed landing gear, the XF3U-1. The SBU-1 went into production and operated successfully from the Navy's carriers.

In 1936 Brewster came out with its first airplane, an XSBA-1 scout bomber mid wing monoplane with a retractable landing gear and a low canopy that must have made for terrible crew visibility. Brewster had little in the way of production facilities and the Naval Aircraft Factory took over limited production of a modified version as the SBN-1 in 1940, the latter being used as operational trainers.

The year 1936 was a big time for new scout bomber models. Curtiss gave up on the high wing monoplane idea as the XSBC-1 and developed via the XSBC-2 model a new production SBC-3 and afterwards an SBC-4 biplane scout bomber with retractable landing gear. These aircraft served into the start of World War II and were among the last of the Navy biplane combat types. In 1936

Grumman also produced a biplane prototype of generally similar configuration to the Curtiss SBC-3/-4 airplanes, but the Navy made this XSBF-1 model one of the very few Grummans not ordered into production. In the same year Vought had two prototype scout bombers in the works, these the XSB2U-1 low folding wing monoplane with retractable landing gear and the XSB3U-1 biplane modification of the SBU-1/-2 production aircraft with retractable landing gear, again showing Navy ambivalence between biplane and monoplane at that time. The monoplane Vought was selected for production as the SB2U-1 through -3 Vindicator used for awhile in the early part of America's participation in World War II.

The first Navy scout bomber destined for true World War II fame was the Douglas SBD- series derived from the Northrop BT-1/XBT-2 airplanes. The SBD- Dauntless flew first in 1940 and was operated through most of the war, being particularly famous for key employment in the 1942 Midway victory as a dive bomber.

In 1941 the Navy had prototypes under development for two new scout/dive bombers, one of which was to supplant the SBD- in wartime operations after a protracted development time. This aircraft was the Curtiss SB2C- Helldiver, sometimes called "The Beast". The other was Brewster's SB2A- Buccaneer and export Bermuda which in total was a disappointment, and was not used operationally by the US Navy.

The last airplane type in the Navy scout bomber category was the attempted Douglas follow-on to the Dauntless, the XSB2D-1. This plane was a large heavy complex design with two rear gun turrets and a retractable tricycle landing gear. Other than two prototypes it was not produced and the design was later simplified into a BTD- bomber torpedo airplane.

As noted above the Navy scout bomber (VSB) category evolved into the VBT and then the VA attack category post-war.

BREWSTER XSBA-1 (Thorell) The XSBA-1 was the first airplane put out by Brewster Aeronautical Corp. after a Navy order for a single prototype in October of 1934. A modern mid-wing two seat monoplane dive bomber of metal construction with space for a 500 pound bomb in an internal bay, it first appeared in April of 1936. Shortly afterwards with cockpit canopies and powerplant modified the plane re-appeared in early 1937. The production contract was given to the Naval Aircraft Factory as the SBN-1. Powered by a Wright Cyclone engine of 950 horsepower the 5740 pound gross weight aircraft had a high speed of 263 mph.

BREWSTER XSBA-1 (H.Thorell) Another view of the single Brewster XSBA-1 prototype after its 1936 modifications shows the details of the new raised canopy for pilot and gunner, Landing gear configuration was generally similar to the later Brewster F2A-/339 Buffalo fighters with wheels fitting into fuselage pockets. The aircraft had a single fixed gun forward and a .30 caliber flexible gun aft. It was felt that Brewster did not have the production facilities to handle a large order.

BREWSTER XSB2A-1 (USN) The single prototype of a new 1000 pound dive bomber for the navy was ordered from Brewster in April of 1939. Shown in a Navy publicity photo of June, 1941 at Newark, NJ. Airport near the big Brewster assembly hangar, the XSB2A-1 was a two place mid-wing monoplane on long landing gear with a dummy rear turret. An actual turret never appeared on the airplane, and after many problems the airplane was finally flown to NAS Patuxent River, Md. on March 9, 1943 as a single seat SB2A-1. Powered by a Wright R-2600 of 1700 horsepower, the approximately 5.5 ton gross weight XSB2A-1 had a high speed of about 310 mph.

BREWSTER SB2A-1 (J.Weathers) A photo of the single prototype after modifications shows the SB2A-1 with a new camouflage paint job, the same large spinner for the Curtiss propeller, new cowl inlets up top, two nose machine guns under the engine, and revised tail surfaces. The plane was operated as a single seater. Wings folded up and over the fuselage; the paint demarcation denotes the fold joint.

BREWSTER SB2A-2 (J.Weathers) The first Navy production model was the SB2A-2 which flew first in September of 1942. The Navy had ordered 202 SB2A-s plus 48 aircraft worth of spare parts; the order was later reduced to 140 planes. However by August, 1943 only about 20 SB2A-2s had been delivered. A total of 80 aircraft were eventually built. The -2s were used as trainers for awhile, but in November, 1943 this operation was discontinued and the planes were used as utility aircraft. In 1944 they went to salvage. Powered by a Wright R-2600 Double Cyclone engine of 1700 horsepower the -2 airplane had a wingspan of 47 feet, weighed up to a little over seven tons loaded, and had a high speed of about 275 mph.

BREWSTER SB2A-3 (J.Weathers) The remainder of the Navy order for 140 SB2A- aircraft, called Buccaneers, was for 60 SB2A-3 versions as shown in the photo. These aircraft were designed for carrier operations, having folding wings and arrestor hooks. They had two .50 caliber nose guns, two .30 caliber wing guns, and twin .30 caliber flexible guns in the rear cockpit. Tests in November of 1943 showed, however, that because of deficiencies in lateral/directional stability and unsatisfactory takeoff characteristics the SB2A-3 was unsuitable for carrier operations.

BREWSTER SB2A-2 (J.Weathers) Flight view of an SB2A-2 airplane illustrates the long "greenhouse" of the Brewster Buccaneer.The big propeller spinner of the prototype has been dispensed with. The test people at NAS Anacostia compared the -2 to the Curtiss SB2C-1 and found the latter was longitudinally unstable in dives but was better in climb than the SB2A-2, though the Brewster was slightly better in level flight. The Curtiss airplane won out for World War II mass production, however.

BREWSTER SB2A-4 (M.Copp) The SB2A-4 aircraft were some (perhaps 105) of the 162 Brewster Bermuda dive bombers ordered by the Netherlands East Indies. By the end of 1942 the Navy had accepted 21 aircraft as -4s with the idea of using them as dive bomber trainers, but the planes exhibited several poor flying quality characterics. In early 1943 some were given to the USMC as night fighter trainers. The Marines encountered propeller, wheel brake, and tail problems. Though some Buccaneers were used as Navy trainers, in 1944 most all the Brewsters were stricken from Naval operations; none ever saw combat.

CURTISS XSBC-1 (Mfr.) A descendent of the two seat XF12C-1 fighter of 1932, the XSBC-1 was originally designated an XS4C-1 scout type and finally the design ended up as the scout bomber shown in a company photo of February 17, 1934. A strut braced high wing two seat monoplane with leading edge slats and trailing edge flaps, the XSBC-1 had retractable landing gear and neatly enclosed cockpits. The plane crashed twice during 1934, the second one finishing it off. Wingspan was 41 feet. The engine was a Wright Cyclone of 725 horsepower providing the aircraft with a high speed of about 220 mph.

CURTISS XSBC-2 (Mfr.) The Navy and Curtiss decided to give up on the high wing monoplane configuration and went to a biplane scout bomber design which was ordered in early 1935 a few months after the demise of the XSBC-1. Shown in a November 29, 1935 company photo, the XSBC-2 was a clean two seat biplane with a novel outboard wing strut arrangement and retractable landing gear. It was powered by a Wright R-1510 Twin Whirlwind engine of 700 horsepower driving a constant speed Curtiss Electric propeller giving the 5800 pound gross weight XSBC-2 a high speed of about 215 mph.

CURTISS XSBC-3 (USN) Late in 1935 the XSBC-2 scout bomber was tested by the Navy and the Wright R-1510 engine was found unsuitable. A Pratt and Whitney R-1535 Twin Wasp Junior of 700 horsepower replaced the Twin Whirlwind and the result was the XSBC-3 shown in this Navy photo of April 17, 1936. Successful tests of this version resulted in a production contract in August. The wingspan was 34 feet, gross weight was about 6500 pounds as a bomber with a 500 pound bomb aboard, and high speed was about 210 to 220 mph depending on the mission load.

CURTISS SBC-3 (P.Bowers) Curtiss obtained a contract in August of 1936 for production SBC-3s and 83 scout bombers were delivered in 1937, the initial aircraft going to Navy Squadron VS-5 aboard carrier Yorktown. The photo shows the SBC-3 of the Yorktown Air Group Commander equipped with a centerline drop tank. Powered with an 825 horsepower Pratt and Whitney R-1535 engine the three ton normal gross weight SBC-3 had a high speed of about 220 mph. Armament consisted of a 500 pound bomb along with a fixed and a flexible .30 caliber machine gun.

CURTISS SBC-3 (W.Larkins) The Navy had three front line squadrons of SBC-3s in the late 1930s but only one of these scout bombers got to the US Marines, the one shown in the photo. A wing bomb rack is apparent as are the ailerons on the upper wing. Flaps were on the lower wings. The final production SBC-3 was converted into an XSBC-4 in 1938.

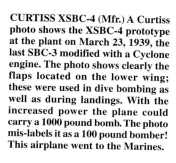

CURTISS XSBC-4 (Mfr.) A Curtiss photo shows the XSBC-4 prototype at the plant on March 23, 1939, the last SBC-3 modified with a Cyclone engine. The photo shows clearly the flaps located on the lower wing; these were used in dive bombing as well as during landings. With the increased power the plane could carry a 1000 pound bomb. The photo mis-labels it as a 100 pound bomber! This airplane went to the Marines.

CURTISS SBC-4 (Mfr.) The change to an SBC-4 model was characterized mainly by use of the higher power single row Wright R-1820 Cyclone engine making the nose of the airplane shorter and larger in diameter, and giving it greater load carrying capability. The first batch of SBC-4s were delivered in 1939; a second batch in 1940. This company photo shows the initial aircraft of the second batch up in cold Buffalo, NY. on February 12, 1940. A 100 pound bomb is carried on the starboard wing rack.

CURTISS SBC-4 (Mfr.) Factory photos (*above, left*) of an SBC-4 before squadron assignment, this aircraft being destined for the reserves. In 1940 a total of 50 SBC-4s were taken back from the Navy and allocated to France, though most all got only to Martinique in the Caribbean where they were interned. Five got to the United Kingdom. Curtiss produced 50 more for the Navy the next year; these had protected fuel tanks. When the US entered World War II SBC-4s were in Navy and Marine service but none got into combat.

CURTISS SBC-4 (P.Bowers) An SBC-4 is shown in early World War II camouflage; note the red center circle is still in the national insignia on the fuselage. Details of landing gear are shown well, and the swinging crutch for a 1000 pound bomb can be seen forward under the fuselage. A tail hook barely protrudes aft. The Wright R-1820 Cyclone developed 950 horsepower and drove a three blade Hamilton propeller. Wingspan was 34 feet, gross weight 6250 pounds, and high speed about 235 mph.

CURTISS XSB2C-1 (Mfr.) A company flight photo shows the XSB2C-1 Helldiver prototype with added vertical fin area. A prototype contract had been awarded in January, 1939 for an XSB2C-1 along with the Brewster XSB2A-1. In December, 1940 a total of 578 production Helldivers were ordered. First XSB2C-1 flight was on December 18, 1940. Extensive testing was done in 1941 and among many changes was an extension of the engine mount by a foot. The Helldiver prototype failed to recover from a zero lift dive on December 21, 1941 and crashed; the pilot bailed out safely. The crash was attributed to horizontal stabilizer failure. Powered with a 1700 horsepower Wright R-2600 engine, wingspan was 49 feet nine inches, gross weight was 10860 pounds, and high speed 322 mph.

CURTISS SB2C-1 (Mfr.) A close-up photo of the first production Helldiver version, the SB2C-1, shows details of engine cowl, wing, and landing gear. The airplane is one of the first 200 produced, much changed from the prototype, but having four .50 caliber guns in the wings which can be seen just outboard of the gear wells. The method of wing fold is illustrated and the wing slats are deployed in low speed position. First production flight tests took place in mid-1942.

CURTISS SB2C-1C (Mfr.) The company photo of August 20, 1943 depicts the prototype SB2C-1C Helldiver which was a converted SB2C-1. The -1C version changed from machine guns to two 20mm cannon in the wings, the C denoting the cannon version. After a long series of delays the Helldiver finally went into combat in November, 1943. The Wright engine still delivered 1700 horsepower maximum, but normal gross weight had risen to 14700 pounds and maximum speed was reduced to about 280 mph. An internal bomb bay accomodated a 1000 pound bomb and there were two wing racks.

CURTISS SB2C-1C (Mfr.) Flight photo of the SB2C-1C shows the outline of the fuselage belly bomb bay, one of the 20mm cannon, and wing racks for bombs or drop tanks. One of many modifications for production was making the tail wheel fixed. The rear seat man normally operated twin .30 caliber or a single .50 caliber weapon. The Helldiver was manufactured at Columbus, Oh.; by the spring of 1944 Curtiss Columbus had produced close to 1000 SB2C-1 Helldivers.

CURTISS SB2C-1 (USN) A Navy photo shows Helldivers on an Essex-class carrier; these airplanes appear to be SB2C-1s with four .50 caliber wing machine guns. Though it had been hoped to get SB2C-1s aboard ship in March of 1943, trials by Squadrons VB-9 and VS-9 in training flights showed innumerable mechanical problems and these squadrons changed back to Douglas SBD-4s in February, 1943. It took a long while to get the Helldiver operational.

CURTISS XSB2C-2 (Edo Corp.) In early 1943 an SB2C-1C Helldiver was converted into a seaplane with twin floats built by the Edo Corporation. Originally envisioned for Marine Corps use, the seaplane version never went into production, and the XSB2C-2 was used for testing of observer twin .30 caliber gun armament at the Aircraft Armament Unit in Norfolk, Va.

CURTISS SB2C-4 (Mfr.) After 1112 SB2C-3 models had been produced in the first half of 1944, these introducing a 1900 horsepower Double Cyclone engine driving a four bladed propeller, the SB2C-4 version was introduced as shown in this company photo showing the wing attachments for a total of eight rockets along with the new perforated wing flaps for dive bombing. This was the major production model with no less than 2045 airplanes delivered. Normal gross weight was 14200 pounds with overload of up to 16600 pounds. High speed was 295 mph at 16700 feet.

CURTISS SB2C-4E (Mfr.) A manufacturer's flight view shows an SB2C-4E version of the -4 Helldiver equipped with additional electronics, including radar. This view illustrates the relatively short tail arm of the Helldiver required by carrier elevator size limits, and the resulting large vertical tail surface area. Early on the tail wheel stayed fixed. Note the carrier arrestor hook position.

CURTISS SB2C-5 (Mfr.) A striking Curtiss photo shows the SB2C-5 version of the Helldiver which was the last in production with a total of 970 aircraft built. These -5 variants appeared in early 1945 with many staying around well postwar. The SB2C-5 had more internal fuel capacity, radar, and less cockpit canopy framing. The Helldivers replaced the Douglas Dauntless in the latter part of World War II.

CURTISS SB2C-5 (Mfr.) Close company flight view of a factory-fresh SB2C-5 Helldiver shows the bullet-shaped power unit cover on the four blade Curtiss Electric propeller, the port 20mm cannon, an unloaded wing store station, and wing mounting posts for rockets along with the perforated wing dive flaps in stowed position. Powered with the 1900 horsepower Wright Double Cyclone engine, maximum gross weight with full bomb load was well over eight tons and maximum speed was 290 mph at a medium altitude.

CURTISS XSB2C-6 (Mfr.) A side view of the final development of the Helldiver, the first of two SB2C-6s converted from earlier versions. Curtiss extended fuselage length about two feet by inserting plugs both forward and aft of the wing. An uprated Wright R-2600 Double Cyclone of 2100 horsepower was installed. The airplane would fit on elevators of the later larger size carriers, but postwar the days of the Helldiver were nearly over and there was no production.

DOUGLAS SBD-1 (Mfr.) The initial version of the famed Douglas Dauntless dive bomber, the SBD-1 is displayed in an early 1941 Douglas (was Northrop) El Segundo plant photo with the airplane, like all -1 versions, destined for the Marines. A tail hook is apparent, however. The SBD-1 evolved directly from the Northrop XBT-2 which was modified into a Dauntless configuration. The first -1s were ordered in late 1940 after first flying in May. A total of 57 were built. Powered by a 1000 horsepower Wright R-1820 Cyclone, the normal gross weight was slightly over four tons and high speed was around 250 mph.

DOUGLAS SBD-2 (Mfr.) The first version of the Dauntless for the Navy carriers, the SBD-2, is shown in a factory photo. Normal gross weight was up almost half a ton, partially due to an almost 50% increase in internal fuel capacity providing well over 300 miles additional range carrying a 1000 pound bomb in its swinging crutch below the fuselage. Eighty-seven -2s were built; high speed was about the same as the -1. The -2 was equipped with three or four .30 caliber guns, two fixed in the nose, and one or two flexible aft.

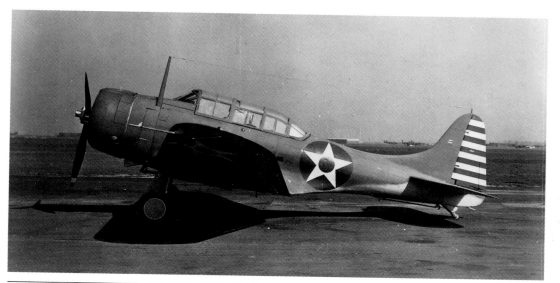

DOUGLAS SBD-3 (Mfr.) A handsome company photo of March 4, 1942, three months into wartime for the US, depicts the SBD-3 Dauntless with the horizontal tail stripes of the period and the red center inside the national insignia star. The centerline bomb crutch can be seen under the fuselage. The nose guns were changed to .50 caliber, the fuel tanks were protected, and armor was added. Gross weight went up to about five tons. SBD-3 production totaled 584 aircraft, not including the SBD-3A/A-24 types for the Army. The -3 airplanes went aboard Pacific fleet carriers, and these were the planes that sealed the victory at Midway for the US in mid-1942.

DOUGLAS SBD-4 (Mfr.) Another Douglas photo shows the SBD-4 Dauntless version, little different from the -3 except for an electrical system change to 24 volts in place of 12. The photo date is November 2, 1942, right at the start of deliveries of 780 -4s, not including Army versions.

DOUGLAS SBD-5 (Mfr.) Although this picture of a new SBD-5 Dauntless was taken at the El Segundo plant on September 18, 1942, major production of this model was at a new Douglas Tulsa, Ok. plant where no less than 2409 SBD-5s were produced for the Navy with an uprated 1200 horsepower Wright Cyclone engine. Production deliveries started in early 1943. Dauntlesses served on US carriers as front line equipment until mid-1944 when they were replaced by Curtiss Helldivers as the standard carrier dive bomber.

DOUGLAS SBD-5 (Mfr.) Quite like the -4 except for engine uprating, an SBD-5 Dauntless is shown in a Navy flight view of July 14, 1943 with a 500 pound bomb on centerline. There were various names for the SBD-; one that the letters stood for Slow But Deadly. The Douglas Tulsa plant produced all the SBD-5s. Performance and weights stayed about the same as earlier wartime versions. With twin .30 caliber flexible guns aft the SBD- was far from a patsy for enemy planes.

DOUGLAS SBD-6 (Mfr.) Apparently Douglas El Segundo , Ca. worked up the Dauntless prototypes and then handed production to the Tulsa plant. Shown with drop tanks on wing store stations is an SBD-6, the final Dauntless version, at El Segundo on March 31, 1944, less than three months before production of the type ended.

DOUGLAS SBD-6 (USN) An SBD-6 Dauntless with additional internal fuel tankage and a Cyclone again uprated to 1350 horsepower for takeoff is shown in December, 1943 flying from the carrier Lexington. The Dauntless would be a front line carrier bomber for at least six months past the date of this picture due to the very late arrival of the Curtiss Helldiver, which arrival was not universally greeted with cheers.

DOUGLAS XSB2D-1 (Mfr.) One of two prototypes of an aircraft meant to supercede the Dauntless, the XSB2D-1 was ordered in mid-1941 but didn't have a first flight until April of 1943. As shown in a company photo four days before first flight the XSB2D-1 was a two seat aircraft with lots of new features, too many no doubt. Noteable were tricycle landing gear, an enclosed belly bomb bay, two remote controlled rear turrets each with a single .50 caliber gun, and two 20mm wing cannon. Using a Wright Duplex Cyclone 18 cylinder engine of 2300 horsepower, the XSB2D-1 had a top speed of about 350 mph.

DOUGLAS XSB2D-1 (Mfr.) Flight test photo of the XSB2D-1 with tufts placed all over the wings inboard shows the gull wing arrangement, doubtless to have the mid-wing carrythrough structure miss the internal bomb bay and still have a reasonably short landing gear. The dorsal fin is truncated to let the dorsal gun swing. Maximum gross weight of the aircraft was near ten tons and wingspan was 45 feet. Since war conditions had changed the Navy soon decided to change the design into a single seat BTD-1 bomber-torpedo aircraft. Given the complications of the XSB2D-1 it was just as well.

GRUMMAN XSBF-1 (H.Thorell) The last scout bomber biplane, the single Grumman XSBF-1 could carry a 500 pound bomb under the fuselage or two 100 pound bombs under the wings. Armament was a fixed .50 caliber, a fixed .30 caliber, and a flexible rear .30 caliber gun. Ordered in March, 1935 and first flown December 24, 1935, it was delivered to the Navy on February 18, 1936. Powered by a Pratt and Whitney 650 horsepower R-1535 Twin Wasp Junior engine driving a Hamilton propeller of eight feet six inch diameter, the wingspan was 34 feet six inches, gross weight 5432 pounds, and high speed 215 mph.

NAVAL AIRCRAFT FACTORY SBN-1 (H.Thorell) The design of the Brewster XSBA-1 scout bomber interested the Navy sufficiently to order in September 1938 30 of the type, not from Brewster whose manufacturing facilities were severely limited, but from the Naval Aircraft Factory at Philadelphia, Pa. The photo shows the mid-wing configuration and the typically Brewster-type landing gear with wheels pocketing into the fuselage upon retraction.

NAVAL AIRCRAFT FACTORY SBN-1 (USN) Flight view of an SBN-1 taken on February 12, 1942, right after US entry into World War II shows the retracted wheels and the perforated wing dive flaps. The 30 aircraft were delivered from late 1940 to mid-1941. A 500 pound bomb could be carried in the internal fuselage bay, and armament consisted of a single .50 caliber fixed gun and a .30 caliber rear flexible weapon. The aircraft were used as trainers for future operational squadrons.

NAVAL AIRCRAFT FACTORY SBN-1 (USN) Another flight view of an SBN-1 on the same date shows the lengthy enclosure over the crew of two. Most all the SBN-1s were delivered by the time this picture was taken. With a Wright R-1820 Cyclone of 950 horsepower and a maximum loaded weight of about 6700 pounds, the SBN-1 had a top speed of about 250 mph. Wingspan was 39 feet.

VOUGHT XSBU-1 (H.Thorell) The XSBU-1 scout bomber by Vought developed from the two seat XF3U-1 fighter of 1933, and this photo shows a partial revision into SBU-1 configuration in the tail but not in the powerplant area. This revision started under an early 1934 Navy contract and the aircraft flew in the spring of that year. The prototype was still in this configuration at NAS Pensacola in 1941 as a utility plane.

VOUGHT XSBU-1 (Mfr.) A manufacturer's photo of September 20, 1935 shows the XSBU-1 still at Vought almost nine months after the company had received a production contract. The aircraft used a Pratt and Whitney Twin Wasp Junior engine of 700 horsepower giving the 5500 pound gross weight 500 pound dive bomber prototype a high speed of about 200 mph with that load aboard. The cockpit enclosure and powerplant installation was never revised to production SBU-1 configuration.

VOUGHT SBU-1 (H.Thorell) An example of the final Navy production biplane scout bomber, the beautiful Vought SBU-1, is shown assigned to NAS Anacostia. A new NACA long chord cowling cleaned up the configuration from that of the prototype, enclosing the uprated Twin Wasp Junior with its long ribbed reduction gear case driving a two blade Hamilton controllable pitch propeller with centrifugal twisting moment counterweights just ahead of the hub.

VOUGHT SBU-1 (P.Bowers) The same Anacostia-based SBU-1, now obsolete, is shown in another view on March 19, 1942. A total of 84 SBU-1s were ordered by the Navy at the start of 1935 and first flight was in August. Delivery took place that year and the next. The SBU-1 served carrier scouting squadrons and some were used in the 1939-40 Neutrality Patrol with the national insignia prominent on the forward fuselage. Span was 33 feet three inches; performance and weights were closely similar to the prototype but production airplanes were slightly heavier and slower.

VOUGHT SBU-1 (Mfr.) One of the SBU-1 aircraft acquired by the Navy was a special "flagplane" for use by high ranking naval officers as pictured in this company photo of the modified scout bomber taken on April 10, 1936. Special features included a non-standard cockpit canopy with built up rear turtledeck and panted landing gear as well as a special pocket for the admiral's stars on the side of the fuselage. The flagplane made a particularly attractive aerial taxi. A tail hook was retained to allow visits to the aircraft carriers.

VOUGHT SBU-2 (P.Bowers) In late 1936 a new minor variant of the SBU-1 with a new model of the R-1535 Twin Wasp Junior engine of 750 horsepower and other minor changes was added to the extent of 40 aircraft. A November, 1942 photo of an SBU-2 is interesting in that it shows the crutch for swinging a 500 pound bomb to clear the propeller stowed under the fuselage.

VOUGHT SBU-2 (USN) Navy flight photo of an SBU-2 shows the lines of one of the cleanest biplanes ever built. These last biplane Voughts were delivered to Naval Reserve squadrons in 1937. The type was used by both Navy and Marine Reserves with two dozen of the SBU-2s delivered to them new; these were at NAS Pensacola at the start of World War II for the US.

VOUGHT XSB2U-1 (Mfr.) A Vought photo taken on January 5, 1936, the day after its first flight, shows the prototype of a new clean-lined monoplane scout bomber, the Vought XSB2U-1. With its long cowled twin row engine, a Pratt and Whitney R-1535 of 700 horsepower, a cantilever low monoplane wing, fully retractable landing gear, and long greenhouse enclosing cockpits for a crew of two, the XSB2U-1 represented the latest, with exception of partial fabric covering on wing and fuselage and of structural makeup, in naval scout bomber design for the mid-1930s. The spread span of the folding wing was 42 feet, maximum gross weight was about 6200 pounds, and high speed was 230 mph.

VOUGHT SB2U-1 (USN) After testing the prototype in the spring of 1936 the Navy gave Vought an order for 54 production SB2U-1 scout bombers in late October, these having a fixed and a flexible .30 caliber gun and the capability of carrying up to a 1000 pound bomb in a centerline swinging crutch. The photo shows the Squadron Commander's plane of Navy VB-3, the High Hats, on December 22, 1937 when the first aircraft was delivered to the carrier Saratoga. The controllable pitch propeller could be put into flat pitch to act as a brake in a dive.

VOUGHT SB2U-1 (UNK.) A flight view of the initial SB2U-1 prior to squadron assignment shows the lean lines of the new monoplane. The SB2U-1 was powered by an 825 horsepower Pratt and Whitney Twin Wasp Junior engine giving a high speed of about 250 mph. Gross weight was 6300 to 7300 pounds. By June, 1940 SB2U-1 and -2 aircraft equipped VB squadrons aboard carriers Lexington, Saratoga, and Ranger. But by the time of US entrance into World War II in December, 1941 the bombing squadrons on Lexington and Saratoga had re-equipped with Dauntlesses. The SB2U-s were assigned for training at Norfolk or used in enlarged squadrons on Ranger and Wasp.

VOUGHT SB2U-2 (H.Thorell) One of 58 SB2U-2 models, named Vindicators at some point, is shown; changes were minor and weights and performance were about the same as those of the SB2U-1. The plane shown was first accepted in Navy squadron VT-3 on Saratoga February 27, 1939, later served in Squadron VS-72 as pictured, and ended up stricken from service in Squadron VS-71 on carrier Wasp on April 30, 1942. Most SB2U-2s ended up in the Navy Training Command by the fall of 1942.

VOUGHT SB2U-3 (USN) A total of 57 SB2U-3 Vindicators with .50 caliber wing guns and self sealing fuel tanks went to the Marines after flying first in January of 1941. The -3 was a little heavier and slower than the earlier models with a top speed of about 240 mph. The aircraft shown in a Navy photo of April 25, 1941 was accepted into Marine Squadron VMS-2 about a month earlier. It ended up being lost in action out of Squadron VMSB-231 during the Japanese December 7, 1941 attack on Pearl Harbor.

VOUGHT XSB2U-3 (Mfr.) A company photo shows the final SB2U-1 airplane revised to an XSB2U-3 for test prior to fitting with floats as a seaplane. The Vindicator landing gear was arranged so wheels twisted to lay flat and flush in the wing undersurface. The two gun ports can be seen in the wing leading edges. Note the large size of the national insignia under the port wing. The -3 was delivered to Anacostia on May 10, 1939.

VOUGHT XSB2U-3 (Mfr.) A twin float seaplane version of the Vindicator was tested in 1939 as shown in the photo, this in common with testing seaplane versions of other Navy aircraft like the TBD-1 and F4F-3. The intended operational use is not clear. Note the ADF loop under the long greenhouse ahead of the rear seat man, the water rudders on the pontoons, and the vertical surface added to counter the directional instability engendered by the pontoon modification.

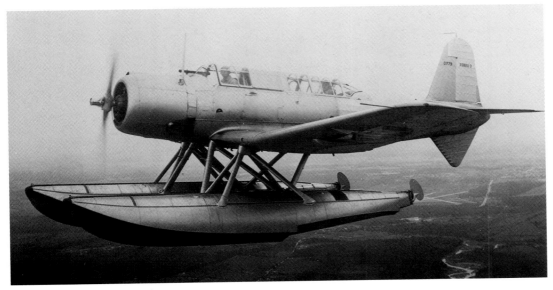

VOUGHT XSB2U-3 (Mfr.) Another photo of the XSB2U-3 seaplane version. The maze of struts supporting the twin floats, a total of 14, is noteable. There were no production seaplanes; the airplane ended up stricken from the Navy on June 26, 1943 from Operational Training Unit One at Jacksonville, Fl. A couple of months later all the Vindicators were gone.

VOUGHT XSB3U-1 (Mfr.) Illustrating the Navy's ambivalence about monoplanes versus biplanes was the Vought XSB3U-1 biplane, this basically an SBU-1 with a narrow tread retractable landing gear. Ordered in February, 1935, about four months after the XSB2U-1 monoplane order, it was tested against the latter airplane prototype the next year. The XSB3U-1 is shown at the Vought factory on February 28, 1936. It was about 15 mph slower than the monoplane with essentially the same powerplant and approximately 300 pounds lighter. Range characteristics were about the same. The monoplane won the day and the XSB3U-1 remained a single prototype.

NAVY SCOUT TRAINER AIRPLANES

For a short period from about 1939 through the late 1940s the Navy used a category of aircraft known as scout trainers (VSN). These airplanes corresponded roughly to the Army basic or advanced trainer types of the same period. There is no indication the Navy ever used the VSN airplanes in a scouting role, so the categorization is a bit of a mystery. It is interesting that one of the most famous US military planes, the North American Texan was placed in this category.

In 1939 North American Aviation produced a new version of their trainer aircraft series as an all metal two seat low wing monoplane with retractable landing gear, the SNJ-1, equivalent to the Army's AT-6, both with the Pratt and Whitney Wasp engine. Hundreds of the SNJ- series through -6 were produced during World War II and employed as Navy Texan trainers. Many still exist today as hobbyist airplanes.

In 1941 just before America's entry into World War II two Navy VSN aircraft types were procured; the first was the Curtiss St. Louis

Division's SNC-1 Falcon, a Navy version of the company Model CW-22 produced also for export. Scores were used as Navy trainers using Wright Whirlwind engines. The second VSN airplane was the Vultee SNV-1 Valiant, a Navy version of the Army BT-13 basic trainer with hundreds procured to train Navy pilots in wartime.

In 1942 the Navy procured its first Beech SNB- Navigator twin engined trainer, this another version of the civil Beech Model 18 light transport. The SNB- was used in various training roles including bombardier training, navigational training, photo planes, and as a utility transport. Hundreds were procured and a large number remained to be updated for further Navy use after the war. Late models were redesignated in 1962 as UC-45Js or RC-45Js.

In the late 1940s North American attempted to produce a successor to the SNJ- Texan as an XSN2J-1 with a more powerful engine and numerous other changes. The prototypes were tail down landing gear aircraft. Later this design was changed to a nose wheel gear arrangement and ultimately the design became the T-28B Trojan Navy trainer.

BEECH SNB-2/TC-45J (R.Besecker) Shown postwar at NAS Lakehurst, NJ. is a TC-45J (after 1962), formerly an SNB-2 Navigator scout trainer. Another of the many versions of the civil Beech 18 twin engine light transport, the SNB-2 was the Navy/Marine equivalent of the Army AT-7. The SNB-1 was the equivalent of the Army AT-11 bombardier trainer. Close to 500 SNB-2s were acquired by the Navy starting in 1942. The SNB-2 carried a crew of five including places for three navigation students.

BEECH SNB-2C/TC-45J (J.Weathers) Another post World War II view of a TC-45J, this at the Cherry Point, NC. Marine Base. The aircraft was originally designated an SNB-2C scout trainer used by the Marine Corps as a navigation trainer and a liason transport. About 375 SNB-2C types were procured and were equivalent to the Army AT-7C. Powered by two Pratt and Whitney R-985 Wasp Junior engines each of 450 horsepower, the 8060 pound gross weight SNB-2C had a high speed of about 225 mph. There also was an ambulance version known as an SNB-2H and an SNB-2P photo version.

BEECH SNB-5 (P.Bowers) The aircraft shown on July 30, 1955 was the last SNB-2C but was sent back to Beech in 1951 for modernization in a program to update powerplants and systems. If this aircraft survived until 1962 it would have been redesignated a TC-45J utility transport. In this case a typical arrangement would be a crew of two and six passengers, though there were several cabin arrangements possible.

BEECH UC-45J (R.Esposito) A handsome 1964 shot of a Navy UC-45J after its redesignation from an SNB-5 trainer two years earlier and modernized in the early 1950s from an original SNB-1 bombardier trainer. Twin Beech 18 aircraft were well known for longevity; this plane was probably built in 1942. Pratt and Whitney Wasp Junior engines were standard. Wingspan was 47 feet eight inches and gross weight around four tons. High speed was slightly over 200 mph.

BEECH SNB-5 (Mfr.) A company flight photo of a late wartime SNB-2 remanufactured and updated into a practically new SNB-5, the picture being taken in the 1950s. A Navy version of the civil Beech 18S, it carries the original number of an SNB-2 on the aircraft. Before World War II ended over 5200 Beech 18S versions were ordered by the Army and Navy in many forms.

BEECH SNB-5P/RC-45J (R.Besecker) Another photo of an ex-SNB-2 wartime Beechcraft shown in the 1960s after having been modernized at the factory first into an SNB-5P photo plane and later redesignated an RC-45J light transport. Original price of a military/naval Twin Beech varied from $51000 to $60000 each. Gross weights of C-45 aircraft varied from 7660 to 8700 pounds. Maximum speed could be as high as 240 mph.

CURTISS SNC-1 (USN) The Navy tested the 1939 prototype of the Curtiss Wright CW-22 and ordered 150 SNC-1 Falcon scout trainer versions the next year in November, 1940. An SNC-1 is shown at NAS Anacostia on May 27, 1941. An all metal low wing two place aircraft with landing gear retracting into bulbous enclosures under the wing, the SNC-1 was powered by a 420 horsepower Wright R-975 Whirlwind engine. The Navy ordered 155 more during wartime, some with an enlarged cockpit enclosure.

CURTISS SNC-1 (USN) A Navy flight view of an SNC-1 Falcon taken near Anacostia on August 13, 1941, this one of the second of three batches procured. The canopy is still on the small side. The Whirlwind engine powered the Falcon to about a 200 mph maximum speed. Wingspan was 35 feet and gross weight about 3700 pounds. Fixed .30 caliber and flexible .30 caliber guns could be installed.

CURTISS SNC-1 (P.Bowers) Several SNC-1 Falcons were purchased by civilians postwar; one of these is shown in a fine runup shot. The Curtiss designation for the Navy production aircraft was CW-22N. Many CW-22s were exported by Curtiss Wright including 50 to Turkey and 36 to the Netherlands East Indies in early World War II. The 1939 prototype was still flying in the late 1960s and perhaps even today.

NORTH AMERICAN SNJ-1 (H.Thorell) The Navy SNJ-1 scout trainer was a version of the Army BC-1 basic combat aircraft. Eighteen of these SNJ-1s were ordered in 1938 with a semi-monococque aft fuselage, a Pratt and Whitney R-1340 of 600 horsepower, and retractable landing gear. Gross weight was about 5220 pounds, span 42 feet seven inches, and the aircraft had a high speed of about 205 mph.

NORTH AMERICAN SNJ-2 (USN) The SNJ-2 was the version sent to Naval Reserve bases. A total of 61 SNJ-2s were ordered in 1939-40. A trim attractive two seater, several were used as command aircraft before the war. A new SNJ-2 is shown at NAS Anacostia, Md. on May 10, 1940. These planes were all metal except for coverings of control surfaces. Again power was provided by a 600 horsepower Pratt and Whitney Wasp engine and weights and performance were similar to the SNJ-1.

NORTH AMERICAN SNJ-2 (USN) A rear view of the SNJ-2 at Anacostia illustrates the curved rudder trailing edge, the last model produced with this feature. As America entered World War II in December, 1941 a total of 57 SNJ-2 aircraft were on hand. Thirty-four of these were at either NAS Anacostia or NAS Pensacola, Fl. The remainder were usually single aircraft used as command planes at Naval Reserve bases.

NORTH AMERICAN SNJ-3 (P.Bowers) The Navy SNJ-3 scout trainer model was the equivalent of the Army AT-6A advanced trainer initially ordered in 1940. The photo shows the new vertical tail design started on this model. A total of almost 300 SNJ-3s were produced, some in California and some in Texas. These aircraft had removeable aluminum fuel tanks. They had a 600 horsepower Pratt and Whitney R-1340 Wasp engine, a span of 42 feet, a gross weight of about 5150 pounds, and a high speed of around 210 mph. In December, 1941 most SNJ-3s were based in Florida at either Jacksonville, Miami, or Pensacola.

NORTH AMERICAN SNJ-4 (R.Besecker) Flight view of the SNJ-4 Navy Texan shows the configuration well. No less than 1425 of these aircraft were built and were generally equivalent to the Army AT-6C Texan. They used the standardized 600 horsepower R-1340-AN-1 Wasp engine. The AT-6C structure was initially redesigned to minimize use of aluminum alloy with many components made of spot-welded low alloy steel with a savings of aluminum of over 1200 pounds. When an aluminum shortage did not occur the more standard construction was then used again. Gross weight was 5300 pounds and high speed hit 208 mph.

NORTH AMERICAN SNJ-4 (USN) Navy Anacostia flight view of an early SNJ-4 Texan is shown. There were complete flight and engine controls in each cockpit. The aircraft could be armed with .30 caliber machine guns, one in the right wing and a flexible gun in the rear cockpit. In Britain this version was known as a Harvard IIA. This was the last model to use a 12 volt electrical system.

NORTH AMERICAN SNJ-5 (J.Weathers) The SNJ-5 Texan was the equivalent of the Army AT-6D model of 1942 and was little different from the SNJ-4 except for installation of a 24 volt electrical system. High speed and gross weight were the same as those figures for the SNJ-4. The aircraft cruised at about 170 mph and landed at 63 mph. Ceiling was about 22000 feet and range on 111 gallons of fuel in the wing center section was 750 miles. Both Navy and Marines used the airplane in many roles.

NORTH AMERICAN SNJ-7 (F.Dean) A civilian ex-SNJ-7 photographed in June of 1978 at New Garden, Pa., one of many purchased in the surplus market and flown by hobbyists. The SNJ-7 came after 931 final production Navy Texans, the SNJ-6 which corresponded to the Army AT-6F. The Texans lasted for many post-war years and the -7 was the result of an in-service modernization program in the early 1950s. The Texans are noted for generating a lot of powerplant noise due to high propeller tip speeds and are today often mistaken for a "World War II fighter" at modern air shows.

NORTH AMERICAN XSN2J-1 (Mfr.) A company photo shows a North American attempt to replace the SNJ- Texan series in the Navy with a more powerful scout trainer, the XSN2J-1 prototype. Retaining a tail wheel landing gear, the two place XSN2J-1 had a 1350 horsepower Wright R-1820 engine driving a three blade Hamilton Standard propeller giving a high speed of over 325 mph.

NORTH AMERICAN XSN2J-1 (H.Borst) A flight view of the XSN2J-1 in the late 1940s looks just like a T-28B trainer of later years except for presence of the fixed tail wheel shown in the picture. The Navy was initially reluctant to use nose wheel landing gear on carriers, thus the tail-down XSN2J-1. It later changed its mind and the XSN2J-1 became a T-28B with retractable nose wheel in place of a tail wheel.

VULTEE SNV-1 (USN) A Navy counterpart of the Army Vultee BT-13A, an SNV-1 Valiant scout trainer is shown at NAS Anacostia on October 28, 1941, one of the first batch of 175 ordered in August of 1940. A total of 1350 SNV-1s were eventually procured. The wingspan was 42 feet two inches and length 28 feet eight inches. The SNV-1s started as trainers at NAS Pensacola, Fl. and Corpus Christi, Tx. Later 650 SNV-2s were procured with 24 volt instead of 12 volt electrical systems.

VULTEE SNV-1 (USN) Another view of the same SNV-1 Valiant at Anacostia in October of 1941, a little over a month before Pearl Harbor. Most served at naval training bases; a few served with the Marines. The all metal fixed gear trainer was sometimes called the "Vibrator".

VULTEE SNV-1 (J.Weathers) A beautifully restored SNV-1 Valiant shown at an airshow in the 1980s displays details of the design including landing gear, wing flaps, and cockpit canopy. Powered by a Pratt and Whitney R-985-AN-1 Wasp Junior engine of 450 horsepower, the 4360 pound gross weight SNV-1 had a high speed of about 165 mph.

NAVY SCOUT OBSERVATION AIRPLANES

The first use by the Navy of the scout observation category occurred in 1935 with the designation of the famous Curtiss SOC-Seagull. Scout observation planes were intended as shipboard catapult aircraft primarily for cruisers, and employed wing folding to minimize hangar space requirements aboard these ships. By 1940 the last of the VSO type were under development, and these barely lasted through World War II when such technical developments as long range search radar aboard ships rendered catapult VSO aircraft obsolete.

The first VSO type was the result of a mid-1930s Navy competition for a new observation amphibian plane. The competition rules were later revised to omit the amphibian capability as too much of a penalty for the aircraft. The winner of the competition was the Curtiss XO3C-1 which for production without amphibian equipment was redesignated as the SOC-1. The folding wing convertable biplane Seagulls became a Navy staple and served both land and ship based through World War II, even replacing their own intended replacement. In 1939 the Naval Aircraft Factory produced a number of duplicates to the SOC-3 as NAF SON-1s. In 1937 a slightly modified Seagull biplane with more flap area was produced as an XSO2C-1, but there was no production of this variant.

A competition for an SOC- replacement resulted in testing in 1939-40 of two high speed scout observation monoplane designs, the Curtiss XSO3C-1 and the Vought XSO2U-1, both convertable from landplane to seaplane configuration and powered by Ranger engines. The Curtiss airplane won the competition and was put into production after many changes and went into wartime service aboard ship for a time. These SO3C- Seagull monoplanes turned out to be disappointingly poor performers such that they were replaced by their predecessors, the SOC-/SON- biplanes.

Later in the war all catapult scout type aircraft, including VSOs were replaced by the new Curtiss high performance SC- scouts which themselves were shortly rendered obsolete in the post-war era.

CURTISS SOC-1 (P.Bowers) The two place SOC-1 Seagull from Curtiss is shown in a handsome view as a landplane for Navy shore base use. After testing the XO3C-1 open cockpit amphibian the Curtiss entry was declared the winner of a competition and evolved into an XSOC-1 seaplane in March of 1935; at the same time the Navy gave Curtiss an order for 135 SOC-1 Seagulls as battleship and cruiser catapult scout observation aircraft. The first SOC-1 was assigned to the light cruiser Marblehead on November 12, 1935. The SOC-1 was a conventional fabric covered biplane with slats and flaps on the upper wing and ailerons on both wings. The wings folded back to minimize space required on Navy capital ships.

CURTISS SOC-1 (H. Thorell) An SOC-1 seaplane from the cruiser Honolulu being launched from NRAB Floyd Bennett Field in New York in 1939 shows the upper wing flaps deflected and elevators down. Over 100 SOC-1s were in fleet service at the time the US entered World War II, and over 50 were still being used midway through the war. Wingspan was 36 feet; they folded to a width of 12 feet three inches. Armament was one fixed and one flexible gun. A bomb rack on each wing could carry a 116 pound bomb or a small depth charge.

CURTISS SOC-1(Mfr.) A Curtiss flight photo shows an SOC-1 Seagull seaplane prior to assignment in a Navy unit. The Seagull was the last of the Navy operational biplanes aside from trainers, serving through the war by replacing its own intended replacement, and even being used in post war years for a short while. Powered by a Pratt and Whitney R-1340 Wasp engine of 550 horsepower and a nine foot diameter two blade Curtiss propeller, the 5000 pound gross weight landplane (5200 pounds as a seaplane) SOC-1 had a high speed of about 160 mph, a little less with floats and a little more as a landplane.

CURTISS SOC-2 (H. Thorell) An SOC-2 landplane from the cruiser Trenton showing unusual vertical tail stripes runs up, probably in 1939. A total of 40 of these Seagull versions were ordered, first as SOC-1s in May of 1936 and later redesignated as SOC-2s. These were delivered with wheeled gear and there were no cowl flaps for the Wasp engines. When equipped with carrier hooks they became SOC-2As. By mid-1943 a total of 17 SOC-2s remained. Gross weight was about 5150 pounds as a landplane.

CURTISS SOC-3 (W.Larkins) This SOC-3 is shown during service with Squadron VO-4 on the battleship USS West Virginia starting in the summer of 1938. A total of 83 SOC-3 Seagulls were procured under a Navy contract of May, 1937; this was the final Curtiss production version. The high aft turtledeck behind the rear cockpit collapsed flat so the observer's .30 caliber gun could be operated. The SOC-3 was a little heavier than earlier models but performance was similar.

CURTISS SOC-3 (Mfr.) A beautiful Curtiss factory photo of December 23, 1937 shows the second SOC-3 aircraft ready for delivery to Squadron VO-2 aboard battleship Oklahoma. Later in wartime battleship SOC-s were replaced by Vought OS2U- Kingfishers. Curtiss delivered the SOC-s as landplanes; they were later fitted with floats as required for shipboard use. Normally the planes were SO (scout observation) types if the wings folded; if wings were fixed they were OS (observation scout). When SOC- wings folded back small panels in the upper wing center section flipped up to give fold clearance.

CURTISS SOC-3 (Thorell) A command SOC-3 Seagull seaplane for use of the Commander in Chief of the US Fleet. The all blue color scheme and the pocket for the CINCUS stars alongside the fuselage near the cockpits shows on this high level taxi. SOC-s were finally replaced very late in the war by Curtiss SC-1s on big combat ships. Many Navy Seagulls were utilized as command aircraft.

CURTISS SOC-4 (H.Thorell) Three Seagull aircraft were sent to the US Coast Guard in 1938 and operated through 1942 designated as V171, V172, and V173 seaplanes though they were delivered as landplanes from Curtiss and could be so operated from USCG stations. They were painted silver all over with the exception of the upper surface of the top wing, and the standard Coast Guard red white and blue rudder markings.

CURTISS SOC-4 (C.Mandrake) A fine flight view of SOC-4 V173 shows the Coast Guard badge on the forward fuselage. One aircraft was based on a Coast Guard cutter operating off the east coast for awhile and the others operated from west coast stations. Later all three went to the Miami, Fl. Station, and later yet they were modified to an SOC-3A configuration with arrester gear for the Navy.

CURTISS XSO2C-1 (H.Thorell) One minor variant of the Seagull was ordered in January, 1937, this an XSO2C-1 with a longer fuselage and flaps located on both upper and lower wings instead of just upper as on SOC-s, and a slightly different version of the Pratt and Whitney Wasp engine. Because of the added fuselage and flaps the empty and gross weight went up about 100 pounds.

CURTISS XSO2C-1 (J.Schneider) Another photo of the single XSO2C-1 prototype variant of the Seagull where the only apparent difference is the lowered flaps on the inboard section of the lower wing as well as flaps down on the upper wing. Like the SOC-s ailerons were on both wings outboard with a drive strut between them.

CURTISS XSO3C-1 (Mfr.) The original landplane version of the new "high speed scout" Seagull meant to replace the SOC- type biplane is shown in a company photo. Powered by a Ranger XV-770 600 horsepower inline 12 cylinder air cooled engine, the landplane version made an initial flight on October 6, 1939 after a Navy prototype contract of May 8, 1938. The XSO3C-1 was in competition with the Vought XSO2U-1.

CURTISS XSO3C-1 (Mfr.) Another company view of the XSO3C-1 Seagull landplane prototype with single strut panted landing gear shown. Also of interest is the extent of wing ailerons and plain flaps. Wing leading edge slats were also incorporated. A mounting ring for a .30 caliber flexible gun installation is barely apparent; the turtledeck section aft of the rear cockpit was collapseable for gun deployment as on the SOC- and SB2C- designs.

CURTISS XSO3C-1 (Mfr.) Curtiss flight view of the XSO3C-1 in initial configuration shows the clean looking engine installation and the panted fixed gear. Also noticeable is the very considerable distance between pilot and observer apparently to get a decent view for the latter. The airplane configuration was shortly to change considerably. Prototype gross weight was about 5500 pounds.

CURTISS XSO3C-1 (Mfr.) The original configuration of the XSO3C-1 seaplane version is shown. Trouble started quickly. With full load the plane would not get off the water until 150 pounds of fuel had been burned off. Then the tail end of the float would drag in the water at high angles of attack. Flight testing showed the need for more vertical tail area and more wing dihedral. Since wing mounting forgings could not easily be changed upswept wingtips were tried and first tested September 11, 1941. The Ranger engine had bearing failure trouble. The single pedestal float mount allowed sufficient deflection in rough water tests to permit the propeller to slice the float. A float cut-out was made and filled with balsa wood and painted gray to match the float.

CURTISS SO3C-1 (Mfr.) An SO3C-1 Seagull runs up its Ranger engine in a company photo at the new Curtiss Wright plant in Columbus, Oh. Major differences from the prototype in engine cowling and dihedraled wingtips are apparent. The wing leading edge shows the slats incorporated. The Ranger V-770 produced 530 horsepower maximum, but was considered an unreliable powerplant and the plane was too heavy, up to 7000 pounds maximum, several hundred pounds over cruiser catapult weight limits.

CURTISS SO3C-1 (Mfr.) A grainy Curtiss photo shows an SO3C-1 Seagull in flight and illustrates the increase in vertical tail area from the prototype. Other models, little different in characteristics, were an SO3C-2 with a tail hook for use on small carriers, and -2C Seamew models for Britain, and an SO3C-3 without a hook. A total of 800 SO3C- types were produced; all were apparently disliked.

CURTISS SO3C-1 (Mfr.) A lineup of production SO3C-1 landplanes is shown at the Curtiss Wright plant in Columbus, Oh. All SO3C- aircraft were delivered as landplanes with floats later fitted as required. Initial SO3C-1s came off the line in the spring of 1942; for some reason production continued through 1943. High speed of an SO3C- was about 180 mph as a landplane and 165 mph as a seaplane.

CURTISS SO3C-1 (USN) A Navy flight photo shows an SO3C-1 Seagull as a floatplane. The increase in size of the cooling air scoop for the engine is apparent. With the crew flying "open cockpit" it is clear the observer's hatch slid forward and carried part of the dorsal fin with it. By early 1944 all SO3C- Seagulls had been removed from front line service; those on cruisers were quickly replaced by their own predecessors, the biplane SOC-s.

NAVAL AIRCRAFT FACTORY SON-1 (H. Thorell) To supplement Curtiss production of Curtiss biplane SOC- Seagulls for cruiser use the Naval Aircraft Factory initiated production of a duplicate SOC-3 labeled an SON-1 shown here in landplane configuration. A total of 44 were produced in the Philadelphia facility in 1939 after an order of June, 1937 making a total of 307 biplanes obtained from Curtiss and the NAF.

NAVAL AIRCRAFT FACTORY SON-1 (H.Thorell) A handsome Navy photo of an NAF SON-1 Seagull landplane on April 3, 1939, probably at NAS Anacostia, Md. prior to an operational squadron assignment. As with the SOC-3 the SON-1 was powered by a Pratt and Whitney R-1340 Wasp engine of 550 horsepower. The landplane version shown weighed about 5730 pounds loaded and had a high speed of 163 mph. Wingspan was 36 feet, and wings folded aft to fit a cruiser hangar.

VOUGHT XSO2U-1 (Mfr.) In 1938 the Navy let a prototype contract for the Vought XSO2U-1 as a competitor to the Curtiss XSO3C-1. The landplane variant shown first appeared for test in July of 1939. The spindly landing gear was less elegant but probably more practical than that of the Curtiss competitor. The XSO2U-1 was powered by the same engine, a Ranger V-770 of 500 horsepower giving the landplane a high speed of about 200 mph. Gross weight was 5500 pounds and wingspan 38 feet two inches.

VOUGHT XSO2U-1 (Mfr.) First flying in late 1939, the XSO2U-1 is shown in its seaplane version for use as a Navy cruiser high speed scout. The wheels are part of the seaplane beaching gear. Dual widely spaced streamlined struts are used to support the main float as opposed to the single pedestal support of the Curtiss plane. Tip floats were supported by a single strut however. Like the Curtiss type pilot and rear seat man were very far apart to prevent the wing from blocking a downward view.

VOUGHT XSO2U-1 (Mfr.) Another view of the XSO2U-1 prototype from Vought as a seaplane with the main float very close to the fuselage. As a seaplane gross weight was about 5500 pounds and high speed 190 mph. The Vought aircraft did not win a production contract; Vought was very much pre-occupied with the XOS2U-1 Kingfisher for battleships and the new XF4U-1 Corsair Navy fighter.

NAVY TORPEDO AIRPLANES

One of the first missions defined for Navy aircraft was carrying and delivering a torpedo against an enemy ship. The airplane was early regarded as a good torpedo delivery system since it was faster and more maneuverable than a submarine or surface ship. The first US Navy torpedo carrying airplanes were Curtiss R-6Ls of 1918; for almost 15 years afterwards the Navy developed a series of specialized single mission torpedo airplanes, both land and carrier based, in the VT class. The VT category faded out in the 1930s with development of the dual mission VTB torpedo bomber.

The Curtiss R-6L, the L designating the Liberty engined version, was selected for US Navy torpedo plane experimental development at the end of World War I in 1918 because of its decent load carrying ability for the time, though this ability was limited to carrying only a 1000 pound torpedo. The next interim type, the first to be used in the new torpedo squadrons, was the NAF PT- based on a combination of Curtiss R-6L and HS-2L components with a Liberty engine.

In 1919 the Navy ordered a few of the Martin twin engined land biplane bombers as MBT torpedo planes because of their ability to carry a larger 1760 pound torpedo. These planes were used in torpedo launching development experiments early on and were later used by the Marines.

In 1921-23 the Navy experimented with several aircraft to be used as torpedo planes; most of the designs were ultimately rejected. One was the Curtiss CT-1 twin float cantilever wing monoplane using two Curtiss engines which never got beyond the prototype stage. Another was a Stout airplane. Still another was the series of clean looking Fokker FT- twin float torpedoplane prototypes with a cantilever monoplane wing and a single Liberty engine. The competition winner was the Douglas DT-1 naval development of the civil Cloudster airplane and the company's first military airplane. A conventional single place (later two place) Liberty engine powered biplane convertible from wheels to twin floats with foldable wings, the DT- series became the standard Navy torpedo plane. The type was also manufactured by the Naval Aircraft Factory and the LWF Engineering Company. An SDW- version for long range scouting was also built by LWF.

In 1927 Boeing produced three big TB-1 prototype torpedo planes with Packard water cooled engines in a design convertable from wheels to floats but did not gain a production order. About the same time Douglas put out a heavier twin radial engined T2D-convertable biplane torpedo plane which could carry a heavy type torpedo and could also operate from an aircraft carrier, but these machines were later used only as patrol planes based on an agreement with the Army on roles and missions. The XT2D-1 had been developed from the experimental Naval Aircraft Factory XTN-1 prototype of 1927 which also used twin radial engines. In the same year the new production torpedo plane was the Martin T3M-1, a three place biplane multi-purpose design based on the earlier Curtiss SC-1 type manufactured by Martin. The T3M- models, -1 and -2, replaced earlier Douglas DT-s and were standard Navy torpedo craft until the early 1930s.

In 1928 the Martin T4M-1 torpedo plane appeared using an air cooled Hornet radial engine and became the new Navy standard with over 100 aiplanes ordered and was used from both land and ship bases. In the following year a near duplicate of the T4M-1 was produced as the Great Lakes TG-1 and TG-2 when the Great Lakes company purchased the design and the plant to build it in from Martin.

In 1930 Martin and the NAF both produced prototypes in the VT category. The Martin XT5M-1 and the NAF XT2N-1 were more dive bombers than torpedo planes however, and very similar, having been built against the same Navy specification. The Martin prototype led to the BM- series dive bomber.

Two final 1931 torpedo planes in the same general biplane configuration were tested by the Navy without any production resulting. The Douglas XT3D-1 prototype was tested and extensively modified into a -2 configuration but ended up only as an engine test bed. Martin came up with the XT6M-1 employing a new metal fuselage, but the biplane era of torpedo planes was near an end, and the next production torpedo carriers would be dual mission all metal monoplanes as VTBs. There were no more VT airplanes.

CURTISS R-6L (R.Besecker) The first Navy aircraft to be tested as a torpedo carrier was the Curtiss R-6L of 1918-19. A development of the basic Curtiss R design, the R-6L shown was one of 40 R-6s equipped with a 400 horsepower Liberty engine in place of the earlier lower power Curtiss engine. This revision enabled the design to carry a 1000 pound torpedo. The R-6L twin float seaplane equipped the first two Navy torpedo squadrons. Wingspan was 57 feet, gross weight about 5500 pounds, and high speed about 95 mph.

BOEING TB-1 (Mfr.) Shown at the Boeing factory on April 15, 1927 after a 1925 order is a Boeing built Navy design for a torpedo plane convertable from wheels to twin floats, the Boeing TB-1. About three weeks later one of the three TB-1s ordered made a first flight. The three place design had folding wings and was intended for basing either on carriers or battleships. The initial aircraft crashed prior to delivery. There was no further production. Powered by a liquid cooled 750 horsepower Packard engine, the big TB-1 had a span of 55 feet, a gross weight of 10700 pounds, and a high speed of about 115 mph.

BOEING TB-1 (Mfr.) The TB-1 is shown in twin float seaplane form on April 26, a day after the previous photo was taken. Two cockpits housed a crew of three; the torpedo officer sat side by side with the pilot just forward of the wing, and a rear gunner was placed aft. The TB-1 was quite similar to the Martin T3M-1 design, but Martin got production contracts and Boeing did not.

CURTISS CT-1 (P.Bowers) One of several torpedo plane designs offered the Navy in 1922 was the Curtiss CT-1, an unusual twin engined cantilever monoplane design with twin booms and tails along with twin floats and a truncated fuselage. The torpedo was carried under the wing center section as shown. The photo was taken on January 7, 1922. The CT-1 had the configuration of the later Lockheed P-38 fighter only as a seaplane!

CURTISS CT-1 (Mfr.) Another view of the CT-1 torpedo plane in the water without a torpedo aboard. The plane was three place with individual cockpits in tandem. Nine CT-1s were ordered but only one was built. Equipped with two Curtiss D-12 engines of 400 horsepower each, wingspan was 62 feet, gross weight 11200 pounds, and high speed about 105 mph. The cantilever wing was unique for an American design of the time.

DOUGLAS DT-1 (Mfr.) A Douglas photo of November 29, 1921 shortly after delivery to the Navy illustrates the first DT-1 as a landplane with a pilot only lugging an 1800 pound torpedo closely hung under the fuselage. Three DT-1s were ordered in the spring of 1921 with design patterned somewhat like the civil Douglas Cloudster, but two were revised to a DT-2 configuration. The biplane wings could be folded back alongside the fuselage. Spread wingspan was 50 feet.

DOUGLAS DT-1 (USN) The same DT-1 single seat aircraft on twin floats is shown inside a hangar. The radiator for the water cooled Liberty engine of 400 horsepower was located alongside the engine compartment. The success of the design led to production torpedo planes. Gross weight as a landplane was about 6200 pounds; the seaplane with a torpedo grossed out at about 7000 pounds. High speed in any case was about 100 mph.

DOUGLAS DT-2 (USN) Tests of the DT-1 resulted in a production order for 38 Douglas DT-2 torpedo planes plus the two other DT-1s converted to DT-2s. In addition 37 DT-2s were built by other manufacturers, the one shown in the photo of August, 1928 with parachutists on the wings built by Lowe Williard and Fowler. The DT-2 became the standard Navy torpedo plane of the mid-1920s. Navy Torpedo Squadron VT-2 got the initial DT-2. Note the Liberty engine getting "air-cooled" as well as water cooled in the August heat! The DT-2 was a two seater and the radiator was placed right on the nose.

DOUGLAS DT-2 (USN) A DT-2 torpedo plane of Squadron VT-1 is shown in the Hawaii area on April 21, 1925. It was not unusual to operate planes with cowl panels mnissing as in this case. The DT-2 seaplane used a 400 horsepower Liberty engine. As a landplane it weighed 6500 pounds loaded; the seaplane version weighed about 800 pounds more. High speed was around 100 mph in either case. The torpedo weighed just under a ton.

DOUGLAS DT-5 (Wright) As with so many production aircraft, experiments were tried using different engines. The photo shows one of two DT-5 aircraft, these originally Naval Aircraft Factory-built DT-4s using a Wright T-2 liquid cooled engine modified later to Wright T-2B geared engines on DT-5s. Note the Lamblin-type side-mounted radiators on the fuselage. The photo shows what a big airplane the DT- type was.

DOUGLAS DT-5 (USN) Another DT-5 photo with the Wright T-2B engine, this in twin float seaplane configuration. The picture was taken in February of 1924. The Wright T-2 Tornado engine was not a big success, and in any case it would not be long before the Navy decided to do away with liquid cooled engines and concentrate on the air cooled variety.

DOUGLAS DT-6 (USN) A Navy photo of February, 1927 shows what is probably an NAF-built DT-2 converted to use a radial engine as a sign of the future. The engine was the 400 horsepower Wright P-1, not a successful radial engine in itself, but gross weight was lower by about 100 pounds than that of the liquid cooled Liberty engined DT-2 with the same horsepower. The high speed did not vary much, about 100 mph.

DOUGLAS/LWF SDW-1 (Wright) A Lowe Willard Fowler-built DT-2 torpedo plane is shown as modified in 1922 into an SDW-1 long range scouting plane, this being one of three done over by the Dayton Wright Company, thus SDW-1 was a really a scout plane by Dayton Wright, and perhaps would better be placed in the scout section. The major change was addition of a large amount of internal fuel as shown by the bulge added under the forward fuselage.

DOUGLAS/LWF SDW-1 (USN) An SDW-1 scout modified from a DT-2 is shown as a twin float seaplane in a 1922 Navy taxi photo. The extra fuel bulge is again shown. The rear cockpit has two men in it. The Liberty 12 engine drove a three bladed propeller. Performance data has not been found.

DOUGLAS XT2D-1 (Mfr.) The Douglas XT2D-1 was unique in being a twin engined landplane that was similar to the Naval Aircraft Factory XTN-1. Both appeared early in 1927 and were designed to carry a 1600 pound torpedo under the fuselage. Looking more like an Army bomber than a Navy torpedo plane, the XT2D-1 had a nose gunner, two tandem cockpits under the wing, and a rear gunner position. The 57 foot span wings could be folded and the plane could be used with twin floats. Powered by Wright R-1750 engines each of 525 horsepower, gross weight was about five tons and high speed about 125 mph.

DOUGLAS T2D-1 (USN) A Navy flight view of one of nine production T2D-1s shows it as a twin float seaplane primarily because the Army got upset about the Navy operating big twin engined landplanes that looked like Army bombers. Rather than conduct more aircraft carrier experiments with the plane the Navy backed off and equipped the type with floats, operating them as patrol planes like the one shown in the photo of April 20, 1929. About 500 pounds heavier than the landplane version, T2D-1 seaplanes had approximately the same performance as the prototype.

DOUGLAS XT3D-1 (Mfr.) A lean looking torpedo plane prototype delivered for test in September of 1931, the XT3D-1 was powered by a 575 horsepower Pratt and Whitney R-1860 Hornet B engine. The aircraft was first ordered in late June of 1930 as a potential replacement for Great Lakes torpedo planes. It was later revised as an XT3D-2. The Hornet engine gave the four ton gross weight aircraft a high speed of about 125 mph.

DOUGLAS XT3D-2 (Mfr.) A January 30, 1933 company photo shows the updated prototype as an XT3D-2 with a more neatly cowled Hornet engine, enclosed crew cockpits, and wheel pants, but performance was only slightly improved and the big torpedo plane remained just a prototype later to be used as an engine test bed.

DOUGLAS XT3D-2 (H.Thorell) Douglas tried hard to keep the new prototype up to date by adding a longer chord engine cowl, wheel pants, and cockpit enclosures, but as seen in the photo the aircraft ended up only as an engine test bed. Though the picture is taken in front of a Wright Aircraft Engines hangar, the installation in the plane is a Pratt and Whitney R-1830 Twin Wasp driving a Hamilton Standard three blade Hydromatic propeller.

FOKKER FT-1 (USN) The Navy tested many aircraft as potential torpedo planes, among them three two place Fokker T3 planes imported from the Dutch plant and equipped with a 450 horsepower Liberty engine. These were designated as FT-1s and as shown in this April, 1923 photo were twin float thick cantilever wing monoplanes. The FT-1s had a gross weight of over 7000 pounds and a high speed of just over 100 mph with a torpedo. No further Fokkers were procured.

FOKKER FT-6 (USN) A revision of the FT-1 is shown in a photo of April 8, 1925. The FT-6 picture gives an excellent view of the thick cantilever Fokker wing. A readily apparent difference from the FT-1 is a new set of floats with a curved instead of a flat upper contour. The radiator for the Liberty engine cooling system is right at the nose.

GREAT LAKES TG-1 (USN) A version of the Martin T4M-1 torpedo plane because Great Lakes Corp. bought the design from Martin along with the manufacturing plant in 1928, the TG-1 was bought by the Navy to the extent of 18 aircraft in a 1929 contract. The first TG-1 is shown as a seaplane at Anacostia, Md. in early 1930. Initial delivery was to Navy squadron VT-2. The aircraft is shown carrying a torpedo and twin Lewis guns are on a Scarff mount at the rear cockpit location. A single .30 caliber fixed wing gun was also fitted. Powered with a Pratt and Whitney R-1690 Hornet engine of 525 horsepower, span was 53 feet, gross weight as a seaplane 8300 pounds, and high speed about 105 mph.

GREAT LAKES TG-2 (USN) The first airplane of 32 TG-2 torpedo types ordered by the Navy from Great Lakes in 1930 is shown in landplane form with a tail hook ready for carrier duty on July 13, 1931. All were delivered that year and served as a standard front line biplane torpedo bomber until replaced by a TBD-1 monoplane. The aircraft had three cockpits with the middle one for the pilot and the front and rear for gunners.

GREAT LAKES TG-2 (USN) Another picture of a Great Lakes TG-2 as a twin float seaplane with a 1000 pound bomb under the fuselage. The photo was taken at NAS Anacostia on August 17, 1931. A single Browning .30 caliber flexible machine gun is located at the rear cockpit. The seaplane version was 500 pounds heavier than the landplane, but high speed was about the same. Wingspan was 53 feet.

GREAT LAKES TG-2 (USN) A fine view of a Great Lakes TG-2 showing the torpedoman viewing area below the engine, the three open cockpits, and landing gear struts arranged so a torpedo could be slung below the fuselage. TG-2s served aboard the carrier Saratoga with Squadron VT-2B starting in 1932 and lasting into 1937. Powered by a Wright R-1820 geared Cyclone engine of 575 horsepower, the 8500 pound gross weight landplane version had a high speed of approximately 125 mph.

MARTIN MBT (USN) A Navy flight photo shows the USN version of the Army's Martin twin engined bomber, this the MBT, meaning Martin Bomber Torpedoplane presumeably. Ten of these large 71 foot five inch wingspan planes were ordered in 1919 and delivered the next year. Most, and eventually all ,of the MBT torpedo types were used by the USMC. Two Liberty engines of 400 horsepower each gave the six ton gross weight biplane a high speed of just over 100 mph.

MARTIN T3M-1 (USN) The initial version of the three place Martin torpedo plane intended for use off an aircraft carrier is shown in a Navy photo of January, 1927, just after the first of two dozen aircraft came into service late in 1926. It could be used as a twin float seaplane as well, and was powered by a liquid cooled Wright T-3 engine of 575 horsepower driving a three bladed propeller. Pilot and torpedoman sat side by side in a forward cockpit and the gunner sat well aft in the rear cockpit as shown. The five ton gross weight aircraft with a 56 foot seven inch wingspan had a high speed of 108 mph. It was the successor to the Martin SC-2.

MARTIN T3M-2 (USN) In 1927 a further development of the Martin torpedo plane came into service with production of no less than 100 aircraft of the T3M-2 type, one of which is shown from Squadron VT-2B on carrier Langley in a Navy photo of October 25, 1927. Other T3M-2s went to the then-new carrier Lexington with Squadron VT-1S. Changes from the -1 included three cockpits in tandem as can be seen in the photo and use of a Packard engine in place of the Wright T-3.

MARTIN T3M-2 (USN) The Martin T3M-2 became the standard torpedo plane of the Navy for both carrier based and land based units. This Navy photo illustrates the twin float seaplane version, an aircraft of Squadron VT-5 in flight on November 17, 1928, about a year and a half after first fleet introduction. A few were left in 1931. The T3M-2 was powered with the liquid cooled Packard 3A-2500 engine of 750 horsepower which gave the five ton gross weight T3M-2 seaplane a high speed of about 110 mph. The landplane was about 500 pounds lighter but a bit slower.

MARTIN T3M-3 (USN) One T3M-2 was converted to use a radial engine since the Navy was converting to radials on all their aircraft to gain simplicity and lighter weight along with a reduction in gunfire vulnerability. The Navy photo of February, 1928 shows the aircraft revised for a Pratt and Whitney R-1690 Hornet radial engine of 525 horsepower.

MARTIN T3M-3 (UNK.) Another view of the test conversion of a T3M-2 into a radial engined T3M-3. This version led directly to the new Hornet powered production T4M-1. For the T3M-3 wingspan was 56 feet seven inches, gross weight was 8300 pounds reflecting the significant weight reduction with respect to the liquid cooled powerplant. Speed was down slightly to about 100 mph with the reduction in engine power.

MARTIN T4M-1 (P.Bowers) As a result of testing the T3M-3 a new XT4M-1 prototype was succeeded by a production batch of T4M-1s totaling 102 aircraft ordered in mid-1927, a large number for the time. Like the T3M-2, the T4M-1 had three cockpits in tandem, one ahead of the wing, one under it, and the third cockpit aft. The fixed landing gear was configured to allow a torpedo to be carried under the belly. Wingspan was 53 feet and length 35 feet seven inches. The aircraft deliveries started in mid-1928 with the first going on board the carrier Saratoga.

MARTIN T4M-1 (USN) A Navy photo of June 26, 1931 shows a Martin T4M-1 twin float seaplane version at NAS Pensacola, Fl. and illustrates the large size of the aircraft next to the sailor shown. T4M-1s were used both in land- and ship-based Navy units. The aircraft was powered by a Pratt and Whitney R-1690 engine of 525 horsepower and the propeller was a two bladed fixed pitch Hamilton type.

MARTIN T4M-1 (USN) Navy flight view of a T4M-1 on July 11, 1930 shows well the locations of the three tandem cockpits. Some planes were used as patrol aircraft; plane number seven of Squadron VP-10 is shown. In 1930 and 1931 T4M-1s were aboard both the Lexington and the Saratoga. T4M-1s had a seaplane gross weight of about 7900 pounds and a high speed of about 110 mph. The landplane was about 500 pounds lighter and slightly faster.

MARTIN XT5M-1 (UNK.) Although the Martin XT5M-1 was designated as a torpedo plane it was also meant more as a dive bomber capable of hauling a 1000 pound bomb. The side view of the aircraft shows two cockpits in an all metal fuselage. Armament was a single fixed .30 caliber fuselage located forward firing gun and a rear flexible gun installation. The aircraft was the fore-runner of the Martin BM-1 dive bomber. These bombers were designed to pull out of a dive with the bomb still attached.

MARTIN XT5M-1 (Mfr.) Another view in March, 1930 of the Martin XT5M-1 with a propeller spinner. The aircraft was very like the Naval Aircraft Factory XT2N-1 since both aircraft were closely designed to the same Navy dive bomber specification. The engine was a Pratt and Whitney R-1690 Hornet of 525 horsepower giving the 5700 pound gross weight XT5M-1 a high speed of about 130 mph. Wingspan was 41 feet.

MARTIN XT6M-1 (USN.) A new prototype torpedo plane with a metal fuselage and metal framed wings was ordered from Martin in mid-1929. First appearing late in 1930, it is shown during testing in a Navy photo of May 5, 1931 carrying a long torpedo and equipped with twin flexible Lewis guns in the rear cockpit on a Scarff ring, Note the corrugated metal covering on the vertical tail fin.

MARTIN XT6M-1 (UNK.) The XT6M-1 was the final torpedo plane prototype and there was no production; the next torpedo type would be a low wing monoplane Douglas TBD-1. The XT6M-1 is shown minus torpedo and rear guns. The plane had a Wright R-1820 engine of 575 horsepower giving the 6840 pound gross weight torpedo plane prototype a high speed of about 125 mph. Wingspan was 42 feet three inches and length was 33feet eight inches.

NAVAL AIRCRAFT FACTORY XTN-1 (USN) An initial experiment in twin engined torpedo aircraft which could be used as a landplane or twin float seaplane as shown in this April 2, 1927 photo; the NAF XTN-1 was generally similar to the Douglas XT2D-1. The big three place biplane had radial engines mounted just over the lower wing. A bombardier or torpedoman viewing station was located low on the nose below the new gunner station. Ordered in May, 1925, the XTN-1 started tests two years later.

NAVAL AIRCRAFT FACTORY XTN-1 (USN) Another view of the XTN-1 on the same April 27 date shows two large windows in the side of the aft fuselage. The XTN-1 stayed a single prototype torpedo plane. Each Wright R-1750 engine delivered 525 horsepower, the span of each wing was 57 feet, and gross weight was five to six tons. High speed came close to 120 mph.

NAVAL AIRCRAFT FACTORY XT2N-1 (USN) A prototype 1000 pound dive bomber like the Martin XT5M-1 with metal construction except for wing covering, the XT2N-1 of 1930 had two fixed .30 caliber guns in the upper wing and two Lewis guns on a rear cockpit Scarff mount along with the ability to dive and pull out with the 1000 pound bomb carried beneath the fuselage. The XT2N-1 remained a prototype, however, and the Martin BM- airplane got the production contracts. The engine was a Wright R-1750 of 525 horsepower, sometimes with a ring cowl, weighed about 5300 pounds loaded, and had a maximum speed of over 140 mph. Wingspan was 41 feet.

NAVY TORPEDO BOMBER AIRPLANES

With the torpedo carrying Navy airplanes of the early 1930s like the Martin T4M-1 and its close brother the Great Lakes TG-1/ -2 rapidly becoming obsolete a replacement was sorely needed. In 1934 a new class of dual mission aircraft was defined, the VTB torpedo bomber, and a Navy competition for a new design in this class was conducted resulting in the TBD-1, the first production torpedo bomber. The VTB class lasted until late in World War II when it was decided to combine the VTB and VSB scout bomber classes into a new VBT bomber torpedo category. This later class lasted only a short time when the functions were combined into a simple attack (VA) category, this latter still being used today.

The winner of the 1934 torpedo bomber competition was the Douglas XTBD-1 Twin Wasp powered all metal power folding low wing monoplane with retractable landing gear carrying a torpedo externally in a semi-submerged belly location. The TBD-1 became the standard Navy torpedo bomber and served into early World War II as the Devastator, but was nearly eliminated in early wartime actions and was out of production. Remaining planes were replaced by the Grumman TBF-1. The other 1934-35 torpedo bomber competitor was the all metal Great Lakes XTBG-1 biplane also with a Pratt and Whitney Twin Wasp and retractable landing gear. The torpedo or bomb load in this type was carried in a closed bomb bay with doors. The advent of the monoplane was at hand, however, and the biplane XTBG-1 lost out to the XTBD-1.

The major torpedo bomber that fought in World War II was the Grumman TBF- Avenger and its similar counterpart from General Motors Eastern Aircraft Division the TBM-. Winning an award in 1940 for XTBF-1 prototype construction the Grumman, with a Wright R-2600 engine, was accompanied by another prototype award to Vought as the XTBU-1 with a Pratt and Whitney Double Wasp R-2800 engine. With an internal bomb or torpedo bay and the same armament the two aircraft were similar in general configuration though the XTBU-1 had a performance advantage. The Avenger was put into production and served in great quantity through World War II and in varied missions well after the war. The XTBU-1 finally got into limited production in 1944 as a Convair TBY-2 Sea Wolf and was just being delivered for squadron service at the end of the war.

In 1945 the Navy tested a Douglas XTB2D-1 torpedo bomber prototype intended for use on the latest large aircraft carriers. A very large and complicated aircraft, the XTB2D-1 was luckily not ordered into production.

The final airplane in the VTB class was a Grumman XTB3F-1 intended as a TBF- follow-on of the late 1940s. After a protracted experimental and test period the design developed into the AF-1 and AF-2 Guardian production aircraft used for ASW missions in the 1950s.

DOUGLAS XTBD-1 (Mfr.) The monoplane winner of a competition for a new torpedo bomber aircraft for the fleet over the biplane Great Lakes XTBG-1, the Douglas XTBD-1 prototype is shown in a company photo of April 22, 1935. A three place all metal low wing aircraft with power wing folding and able to carry a torpedo or a 1000 pound bomb, the XTBD-1 had a first flight on April 15, 1935 and was shortly delivered to the Navy.

DOUGLAS XTBD-1 (USN) Flight view of the XTBD-1 prototype shows a clean low wing layout that was very modern for 1935, particularly when compared with previous torpedo planes. The torpedo was carried semi-submerged in the belly. Wingspan was 50 feet spread and about half that folded. The 800 horsepower Pratt and Whitney Twin Wasp engine gave the 8500 pound gross weight XTBD-1 a maximum speed of 205 mph.

DOUGLAS TBD-1 (Mfr.) Based on successful testing of the XTBD-1 prototype Douglas obtained the very large order for that time of 114 production TBD-1s in early 1936. The Douglas photo shows the first aircraft at the factory on June 28, 1937 ready for delivery. In October Navy Squadron VT-3 got the last production aircraft. Production changes were an uprated 900 horsepower R-1830 Twin Wasp engine, a five ton gross weight, a revised oil cooler installation, and a large new canopy over the cockpits. This aircraft was used later as a TBD-1A test plane on twin seaplane floats in 1939.

DOUGLAS TBD-1 (Mfr.) An interesting photo shows a TBD-1 of Squadron VT-3 from the carrier Saratoga preparing to land at NAS Floyd Bennett Field in NY. The gear is down and all three crewmen have hatches open. The center crewman is the torpedo officer; he had a belly sighting station. The single .30 caliber fixed gun was at the engine cowl. The TBD-1 became the Navy's standard torpedo bomber with all delivered by late 1939. Performance was about the same as the prototype.

DOUGLAS TBD-1 (J.Weathers) The photo shows how the TBD-1, now called the Devastator, folded its wings up over the fuselage. The landing gear retracted straight back with much of the wheel protruding below wing skin line, Note the oil cooler just inboard of the starboard gear and the fact that wing metal skinning was corrugated. Though initially achieving some success in early 1942 the slow TBD-1s were slaughtered at Midway in the middle of 1942. Soon the type was retired from first line service and was replaced by the Grumman TBF-1.

DOUGLAS XTB2D-1 (Mfr.) Douglas really went overboard and designed a monster new torpedo bomber intended for the latest large carriers planned in World War II, the Midway class. This was the XTB2D-1 with no less than a 70 foot wingspan and weighing as much as 17 tons loaded. Ordered in late 1943, the single prototype is shown in a Douglas El Segundo Ca. photo of February 26, 1945 shortly before delivery. The aircraft had more than enough of most everything, including a tricycle landing gear, a dual rotation propeller, and a 28 cylinder four row Wasp Major engine.

DOUGLAS XTB2D-1 (Mfr.) Another view of the XTB2D-1 prototype on the same date shows its imposing 46 foot length. Two torpedoes could be carried. The design had two remotely controlled turrets each with twin .50 caliber guns along with four fixed heavy machine guns in the wings. The Pratt and Whitney Wasp Major engine developed a maximum of 3000 horsepower and powered the plane to a high speed of about 335 mph. Probably luckily only one prototype was built.

GRUMMAN TBF-1 (Mfr.) An early model of what turned out to be the standard US Navy torpedo bomber of World War II starting about six months after Pearl Harbor is shown. The Grumman TBF-1 Avenger started with two XTBF-1 prototypes ordered in early 1940 to replace the now unsatisfactory TBD-1 Devastators. First XTBF-1 flight was in August of 1941, but the first plane crashed on November 28, 1941 with two crewmen bailing out. The second XTBF-1 was delivered on December 15, 1941 just a week after Pearl Harbor. The photo shows the rear turret with a single .50 caliber gun and a ventral position for a flexible .30 caliber weapon. A single fixed .30 caliber gun was in the cowl. The three place aircraft carried a torpedo, bombs, or mines in an enclosed fuselage belly bay.

GRUMMAN TBF-1 (Mfr.) Flight view of the bulky eight ton TBF-1 Avenger shows the outline of the belly torpedo bay doors. A total of no less than 2291 TBF-1 and TBF-1C Avengers were produced between January, 1942 and the last day of 1943 at Grumman before the pressure of Grumman fighter work forced production to be turned over to General Motors. Powered by a Wright R-2600 Double Cyclone engine of 1700 horsepower driving a 13 foot diameter propeller, the TBF-1 had a high speed of about 270 mph.

GRUMMAN TBF-1 (USN) Two TBF-1 Avenger torpedo planes make a practice run with one torpedo yet to enter the water in an October 29, 1943 wartime photo. The Avengers were America's only torpedo bomber in the war and were also used as bombers, scouts, and for mine laying. With depth charges they became ASW aircraft. Weapons bay doors are fully open on the planes shown. The TBF-1 weighed about 13700 to 16000 pounds loaded depending on the mission.

GRUMMAN TBF-1C (UNK.) On display as a memorial of World War II, a Grumman TBF-1C of the Royal New Zealand Air Forces is one of 42 sent to that country in 1943. The -1 and -1C had engine cowl flaps only on the upper cowl section as shown. The photo indicates how the Avenger landing gear swung outboard to retract into the wing. Grumman built 764 TBF-1Cs. They had a fixed .50 caliber gun in each wing, a .50 caliber dorsal turret gun, and a .30 caliber ventral gun. Normal gross weight was 16400 pounds; high speed was 257 mph.

GRUMMAN XTBF-3 (USN) After Grumman built one XTBF-2 prototype with a first flight on May 1, 1942 using a Wright R-2600 engine model with a two stage supercharger, they continued with one or two XTBF-3 prototypes as shown in a Navy NATC Patuxent River, Md. photo of June 8, 1943. Grumman records the first flight of an XTBF-3 as June 30, 1943. It was the prototype for the General Motors Eastern Aircraft Division TBM-3 production aircraft. The R-2600 engine was uprated to 1900 horsepower. High speed came up to 267 mph. Modifications were made to cowl flaps and landing gear also.

GRUMMAN XTB3F-1 (Mfr.) After an abortive too heavy twin engined XTB2F-1 design was discarded the XTB3F-1 shown was initiated as a TBF- follow-on with a side by side two man crew in a single engine monoplane with a weapons bay for two torpedoes or two 2000 pound bombs. Power was supplied by a Pratt and Whitney R-2800 Double Wasp engine of 2000 horsepower and a Westinghouse jet engine of 1600 pounds thrust. The XTB3F-1 prototype, "Fertile Myrtle", had the jet installed but it was not used in flight test. It was later removed and the wing leading edge air inlets covered over.

GRUMMAN XTB3F-1S (Mfr.) A Grumman photo shows aircraft number 90506 as an ASW hunter aircraft XTB3F-1S flown by test pilot Corwin Meyer in late 1948. The big belly radome housed the search radar antenna. This prototype was tested until 1950 when it was stricken from Navy roles. The other ASW killer aircraft prototype, an XTB3F-1S, crashed in October, 1949. The XTB3F-1 prototypes were used in development of the AF-1 and AF-2 Guardian ASW airplanes.

GREAT LAKES XTBG-1 (USN) The last of the biplane torpedo type aircraft and a biplane competitor to the new Douglas XTBD-1 monoplane, the XTBG-1 from Great Lakes is shown in a Navy photo of August 24, 1935, just four days after delivery. Powered by a Pratt and Whitney R-1830 Twin Wasp engine of 800 horsepower, the 9275 pound maximum gross weight XTBG-1 had a high speed of 185 mph. With a torpedo it was several miles per hour slower.

GREAT LAKES XTBG-1 (USN) Another view of the XTBG-1 prototype on August 23, 1935 shows the bulky all metal fuselage with retractable landing gear. The forward cockpit and the viewing area just forward of the landing gear was for the torpedoman. Pilot and rear gunner were enclosed further aft. Span of the metal framed fabric covered non-folding wings was 42 feet. The design lost out to the TBD-1 monoplane and only the prototype was built.

Eastern TBM-3 (USN) Eastern Aircraft Division of General Motors took over Grumman Avenger production so Grumman could concentrate on new fighter production starting in late 1942. An Eastern TBM-3 developed from the Grumman XTBF-3 prototype, one of over 4600 aircraft, is shown in a March 17, 1945 Navy photo in landing configuration taking a wave-off from Escort Carrier CVE-93. The mission was probably ASW.

EASTERN TBM-3 (USMC) The Marines also used the Avenger as shown in this USMC photo of a TBM-3 at a Pacific Island base in World War II. The double sectional weapons bay doors are open. Like the prototypes the TBM-3 used a 1900 horsepower R-2600 Double Cyclone engine giving a top speed of 267 mph. Gross weight went up to about nine tons and spread wingspan was 54 feet two inches. There was a great deal of space inside the fuselage of this big airplane.

EASTERN TBM-3 (S.Orr) A fine taxi photo of a restored TBM-3 taken at a Lawton, Ok. airshow in September of 1983 shows how the Avenger's wing folded leading edge down around a canted hinge so the outer panels laid back aft. A powered fold system was employed. The weapons bay could accomodate a torpedo or a 1000 pound bomb among other loads.

EASTERN TBM-3D (USN) At least nine different versions of the TBM-3 were made for various missions. The basic -3 had provision for wing drop tanks, had more armor than the -1, and two .50 caliber fixed wing guns along with provisions for HVAR rockets. The TBM-3D version was equipped with an APS-6 radar on the starboard wing as shown in this Navy photo of January 11, 1945.

EASTERN TBM-3E (R.Besecker) At the end of World War II and for some time post-war various mission-oriented TBM-3 versions were employed, one of these the TBM-3E shown in the photo. This version was intended for ASW work based on the increased post-war emphasis put on such versions by the Navy. An APS-4 radar set was carried in the TBM-3E for such a purpose. The aircraft carried a crew of three, pilot, gunner, and radar operator.

EASTERN TBM-3E (R.Besecker) Another view of an Eastern TBM-3E Avenger taken at San Diego on January 3, 1952 shows the aircraft with open weapons bay doors ready to accept a bomb load of over 2000 pounds. The APS-4 radar pod is mounted under the starboard wing. Also shown are the underwing standoffs for mounting HVAR rockets. The 1900 horsepower Wright engine gave the 14160 pound normal gross weight TBM-3E a high speed of 276 mph. Armament was the standard two .50 caliber wing guns, a single dorsal .50 caliber weapon, and a .30 caliber ventral gun.

EASTERN TBM-3S2 (AAHS) Avenger aircraft were employed more and more in anti-submarine missions post-war and were put together in two airplane hunter-killer teams. Shown is the TBM-3S2 killer/strike part of the duo with open weapons bay, radar pod, and rocket mounts at Dallas Tx. on January 23, 1953.

EASTERN TBM-3W2 (AAHS) The other part of the ASW hunter/killer duo was the TBM-3W2 Avenger used for the search part of the mission with the big fuselage belly radome for an APS-20 search radar. The rear turret was removed and the canopy faired down. Additional vertical tail surfaces were provided as shown to counter the directional instability effects of the radome forward. The TBM-3 left front line service in mid-1954; the photo was taken on August 5, 1954 with the aircraft operated in the NAS Willow Grove, Pa. Reserves.

EASTERN XTBM-4 (P.Bowers) A 1945 version of the Avenger was to have been the TBM-4 with airframe updates but only prototypes were built with one of the three shown in this fine photo of the aircraft at NATC Patuxent River, Md, the FT on the fuselage standing for flight test section.

VOUGHT XTBU-1 (Mfr.) The XTBU-1 was ordered by the Navy from Vought as a torpedo bomber backup to the XTBF-1 in the spring of 1940 just a few days after the latter aircraft was ordered. About the same size as the Grumman but powered by a Pratt and Whitney Double Wasp engine of 2000 horsepower, the XTBU-1 was arranged similarly as shown in the company photo and armed with a fixed cowl gun and dorsal and ventral weapons. A belly weapons bay enclosed a torpedo or bombs.

VOUGHT XTBU-1 (Mfr.) Another view of the single XTBU-1 prototype shows the clean cowl enclosing the Double Wasp engine and the retractable landing gear which went aft up into the wing with the wheel twisting to lie flat in typical Vought design fashion at that time. The XTBU-1 had its initial flight in December, 1941 just after the Pearl Harbor attack. Wingspan was 57 feet two inches. Wings folded up over the fuselage; the fold line is shown by the paint demarcation as opposed to the aft folding of the TBF-1 wing.

VOUGHT XTBU-1 (Mfr.) Called the Sea wolf, the backup XTBU-1 torpedo bomber is shown in a company flight photo on March 16, 1942 with torpedo bay doors open and the rear dorsal power gun turret with a .50 caliber gun ready to operate after hatch sections have been moved forward and aft in a somewhat complex appearing arrangement. The 2000 horsepower Double Wasp gave the Sea Wolf a high speed of well over 300 mph. Normal gross weight was about 16200 pounds. Vought was busy with OS2U-1 and F4U-1 Navy aircraft production, and had no facilities for TBU-1 production.

CONVAIR TBY-2 (Mfr.) The XTBU-1 Sea Wolf was to be mass produced to the extent of over 1000 aircraft in a new Convair plant located in Allentown, Pa. A Vought photo of September 7, 1944 shows an early TBY-2 as the aircraft was now known using augmented armament and equipment. Production deliveries started two months later, but the program was late and only 180 planes were put out. The TBY-2 was somewhat heavier and a little slower than the prototype. With production of the Avenger by Eastern/GM the Sea Wolf was not needed and was not used operationally.

NAVY TRAINING AIRPLANES

A short while after World War II the Navy changed the designation of training aircraft from the VN category to VT, this for all kinds of trainers, large and small. From approximately 1947 until 1962 the "old" Navy designation system which incorporated the airplane manufacturer's assigned letter after the T for trainer was used. In 1962 a major revision of aircraft designations was incorporated which in essence folded new and in service Navy aircraft in with those of the USAF, and dropped the manufacturer's letter which never had been used by the USAAF or USAF.

The first Navy trainer in the new T system was the 1948 Lockheed TV- version of the USAF F-80 single seat jet Shooting Star (TV-1) and the T-33 T-bird two seater (TV-2). In 1962 the Navy T-birds became T-33Bs in the new system of designations. Experience with these aircraft later dictated Navy requirements for a much revised later two seat jet trainer, and Lockheed came out in 1955 with the T2V-1 Seastar, later redesignated a T-1A in 1962. The airplane could be used on land or on carrier decks.

In 1952 a new Navy piston engined two seat trainer appeared as a revision of the earlier North American XSN2J-1 scout trainer. A primary change was substitution of a tricycle landing gear for the tail down configuration of the XSN2J-1 prototype. The new aircraft was the T-28B/C produced in quantity both as a land and carrier compatible aircraft. A substantial number of T-28B or T-28C Trojans were for many years used in Navy training programs. In

the same year of 1952 a number of twin piston engine Convair T-29 transports were loaned to the Navy by the USAF without a designation change and used as navigation trainers.

The year 1955 saw Beech delivering T-34B Mentor two place light piston engined primary trainer aircraft to the Navy, and a year later Temco built a few two place light single jet engined primary trainers as TT-1 Pintos for a Navy experiment in using jet aircraft right from the start in flight training. That plan was not implemented. Beech later built a number of T-34C turboprop powered versions of the Mentor for the Navy.

In 1958 North American started deliveries of a single jet powered two place T2J-1 (later in 1962 a T-2A) Buckeye trainer used in the Naval Air Training Command. It was later considered that a twin engined version of this trainer would be more appropriate for the purpose, and the later T-2B and T-2C models, starting in 1965, were delivered equipped with two Pratt and Whitney and two General Electric jets respectively.

The year 1960 saw the start of deliveries of a few navalized versions of the North American Sabreliner in several models for light personnel and cargo transport as well as for training of radar operators. Later several off the shelf civil aircraft were ordered as T-44s from Beech. These were King Air turboprop airplanes used in various training roles.

One of the latest trainers is the turbofan-powered McDonnell Douglas two seat T-45A Goshawk, which came into use in 1992.

TEMCO TT-1 (P.Bowers) Post World War II the Navy became interested in the possibility of a jet powered primary trainer to prepare new pilots for naval jet combat aircraft. A prototype of the light two place single engine Texas Engineering and Manufacturing Company TT-1 Pinto is shown. Developed privately as can be seen by the civil registration number still on the tail, the TT-1 flew initially in early 1956 and was submitted for Navy testing.

TEMCO TT-1 (USN) On the basis of prototype tests of the TT-1 Pinto by the Navy a total of 14 production aircraft were procured and these were delivered in 1957. Six of these trainers are shown in the photo flying from NAS Saufley Field in July of 1957. They were used to assess the practicality of use for primary training. No more were ordered and the Navy went back to piston engined aircraft for this purpose. The 920 pound jet thrust Continental J69 engine pushed the 4400 pound gross weight Pinto aircraft to a high speed of 325 mph.

LOCKHEED TV-1 (F.Dean) The Navy procured 50 Lockheed F- 80C single place Air Force jet fighters in 1948 to provide jet flying experience for Navy pilots in a period when they had no comparable production aircraft. Procured as TO-1 trainers they became TV-1s when the Lockheed company designator letter was changed from O to V. The USMC used a few as fighters. Powered by an Allison J33 jet engine of 5400 pounds thrust the eight ton gross weight TV-1 had a high speed of 580 mph. The TV-1 pictured was on static display at NAS Willow Grove, Pa. in May of 1977.

LOCKHEED TV-2/T-33B (R.Besecker) Like the US Air Force, the Navy made use of a large number of two place jet trainer versions of the Lockheed Shooting Star, the ubiquitous "T-bird" of the 1950s and 1960s. A total of 699 were procured as TO-2s, later TV-2s, and still later in 1962 and after T-33Bs. The T-33B shown in a handsome photo was a 1953 aircraft photographed still being used in November of 1965. The aircraft used a 5200 pound jet thrust Allison J33 engine, had a gross weight of six tons, and a high speed of about 550 mph at altitude.

LOCKHEED T2V-1/T-1A (R.Besecker) In the early 1950s Lockheed developed a major modification of the TV-2 "T-Bird" specifically as a naval jet trainer both for use on land and on aircraft carriers. New fuselage contouring raised the rear seat man to provide a better view, and wings were modified to reduce landing speed. The T2V-1 Seastar, 149 of which were procured starting in the mid-1950s, had its designation changed in 1962 to T-1A. The fine photo of a Reserves T2V-1/T-1A was taken in May of 1962.

LOCKHEED T2V-1/T-1A (R.Besecker) Showing a general resemblance to a T-33, a T-1A is shown in a photo of September 11, 1964. These Seastar aircraft were equipped with tailhooks and so could be accomodated aboard aircraft carriers. A blown flap system assisted in reducing landing speed. Wingtip fuel tanks were a standard fixture. Wingspan was 42 feet ten inches and length 38 feet six inches.

LOCKHEED T2V-1/T-1A (USN) Four T2V-1 Seastar two place jet trainers fly in echelon over an aircraft carrier reflecting their carrier compatibility in a Navy photo of June 13, 1959 near NAS Pensacola, Fl. The 6100 pound jet thrust Allison J33 engine gave the eight ton gross weight Seastar a maximum speed of 580 mph at high altitude. Range was just under 1000 miles.

NORTH AMERICAN YT2J-1 (P.Bowers) After a trainer design competition North American Columbus, Oh. was awarded a 1956 development contract for an XT2J-1 Navy jet aircraft. A single engine tandem two seat mid-wing monoplane, the first of two prototypes flew initially on January 31, 1958 using a Westinghouse J34 engine of 3400 pounds thrust. The plane was to be used from early flight training right on to carrier indoctrination. The single engine plane is shown during test at NATC Patuxent River, Md.

NORTH AMERICAN T2J-1/T-2A (USN) Flight view of the T2J-1 Buckeye redesignated a T-2A in 1962 shows a trim carrier hook equipped monoplane jet trainer. Production from 1958 through January, 1961 ran to a total of 217 single engined T-2As. Some planes were used by the Naval Air Training Command at NATC Pensacola, Fl. until early 1973 when they were replaced by later models T-2B and T-2C. Other T2J-1s were used in training squadrons at NAAS Meridian, Mi.

NORTH AMERICAN T-2C (R.Besecker) A May, 1973 photo shows a T-2C Buckeye jet trainer. After a total of 97 T-2B twin engined versions were built for the Naval Air Training Program from May, 1965 to February, 1969 using Pratt and Whitney J60 engines each of 3000 pounds thrust, North American started a similar T-2C Buckeye in production late in 1968 using two General Electric J85 jet engines. The first production T-2C flew on December 10, 1968. Well over 200 T-2Cs were built.

NORTH AMERICAN T-2C (J.Weathers) A fine photo of a twin engined T-2C Buckeye trainer of Squadron VT-23 from Kingsville Tx. with canopy raised is shown on February 10,1974. The T-2C employed General Electric J85 engines each of 2950 pounds of thrust. Fuel was carried in a tank over the engines (387 gallons); two internal wing tanks carried 100 gallons, and external wingtip tanks each carried 102 gallons. Maximum takeoff gross weight was 13180 pounds and high speed was 522 mph at 25000 feet.

NORTH AMERICAN T-28B (N.Taylor) In 1952 the Navy, after having run tests on USAF T-28A trainers and endeavoring to attain some trainer standardization, ordered a modified version as a T-28B with an uprated powerplant, new cockpit canopy, an air brake under the fuselage, all training armament external with underwing mounts, and new radio and electrical systems. A total of 489 T-28B Trojans were procured. A fine T-28B photo shows a Trojan visiting Kelly AFB, Tx. on July 7, 1972.

NORTH AMERICAN T-28B (P.Bowers) A dynamic runup photo shows a T-28B Trojan to good advantage. The 1425 horsepower Wright R-1820 Cyclone is driving a three blade nine foot four inch diameter Hamilton Standard Hydromatic propeller. The T-28B had the performance of an early World War II fighter plane with a high speed of 346 mph and a cruise speed of 310 mph at altitude. Maximum takeoff gross weight was 8250 pounds, initial rate of climb 3800 feet per minute, and the service ceiling was 37000 feet.

NORTH AMERICAN T-28B (R.Dean) A T-28B Trojan trainer taxis in at a 1970s air show with the crew of two having slid hatches back. The propeller thrust line has a noticeable negative tilt, this to balance out the value of the one-per-revolution propeller vibration excitation factor between flight conditions. Provision could be made for carrying bombs, 2.25 inch SCA rockets, and .50 caliber gun pods underwing. Wingspan was 40 feet seven inches.

NORTH AMERICAN T-28C (USN) The T-28C model was a T-28B equipped with a tail hook as shown on this aircraft of Squadron VT-1 landing with hook deployed to catch a wire. The first T-28C was flown on September 19, 1955. Production ended in the fall of 1957 after 299 of this version had been built. A total of 1948 Trojans were built for the Army (T-28A) and Navy (T-28B and C). Cockpits had duplicated flight controls and instruments for training.

CONVAIR T-29B (UNK.) The photo shows one of a few Convair T-29Bs loaned by the USAF to the Navy, this one used by Squadron VT-29. The USAF serial number is still on the tail. The first T-29B flew in mid-1952 and 105 planes were built, most all used by the USAF. The T-29B had a pressurized cabin, a gross weight of 43575 pounds, and was powered by two 2500 horsepower Pratt and Whitney R-2800 engines giving a high speed of 300 mph. It was normally used as a navigation trainer.

CONVAIR T-29C (UNK.) Another Convair-Liner loan to the Navy from the USAF, this similar to the T-29B but powered with a new Double Wasp engine variant. First flight was on July 28, 1953. Accomodation was made for 14 navigation students and instructors. A total of 119 aircraft were built; the number loaned to the Navy by the USAF is not known.

NORTH AMERICAN T-39B (J.Weathers) A Navy T-39B variant of the North American Sabreliner, the civil prototype of which first flew in September of 1958 using two General Electric J85 jet engines. Shown is one of the early versions done up in markings of the 1976 Bicentennial with two windows per side. The aircraft is from Navy Squadron VT-10 based at Sherman Field, NAS Pensacola, Fl. Note the air brake open beneath the fuselage.

NORTH AMERICAN CT-39C (J.Weathers) A VIP transport version of the Sabreliner with five windows per side, a naval variant of the civil Series 60, the fuselage having been lengthened by three feet two inches over earlier models. There was accomodation in the standard model for a crew of two and ten passengers. The photo was taken at NAS New Orleans on September 14, 1980. Maximum takeoff gross weight was about ten tons.

NORTH AMERICAN CT-39G (J.Weathers) A Navy CT-39G used by the Marines at El Toro is shown visiting NAS New Orleans on June 21, 1978. The CT-39s were procured in small numbers and in various configurations. Initially 42 T-39D versions of the civil Series 40 were procured in the early 1960s as trainers for Navy radar operators. Then the Navy acquired nine CT-39E versions of the Series 40 for rapid response of high priority passengers and cargo in the early 1970s.

BEECH T-34B (R.Stuckey) After testing the Beech civil Model 45 in 1953, and after the USAF procured over 300 T-34A Mentors, the Navy ordered a T-34B version in June of 1954 for primary training duty. The piston engined tricycle gear equipped Navy Mentor is shown in the photo as one of the last built; a total of 423 were procured through October, 1957 when production of the Navy version stopped.

BEECH T-34B (Mfr.) The first Navy T-34B is shown in a 1954 flight photo. A conventional two seat all metal low wing primary trainer, the Mentor was powered with a 225 horsepower Continental O-470 six cylinder horizontally opposed air cooled engine driving a Beech seven foot diameter constant speed propeller. Wingspan was 32 feet ten inches. The gross weight was 2985 pounds and the T-34B had a high speed of 188 mph.

BEECH T-34C (Mfr.) To inject turbine power into Navy primary flight training a follow-on turboprop version of the Mentor was ordered as shown in this photo of two early aircraft. Substantial changes are those in the powerplant area and revised vertical and horizontal tail surfaces. The T-34C aircraft are currently used as primary trainers.

BEECH T-44A (F.Dean) A photo taken on September 3, 1979 at NAS Willow Grove, Pa. shows a Navy T-44A training version of the civil Beech King Air Model 90, a pressurized six to ten place aircraft powered by two Pratt and Whitney PT6A turboprop engines each of 550 equivalent shaft horsepower driving Hartzell constant speed fully feathering propellers. The T-44A has a gross weight of about 9650 pounds and a wingspan of 50 feet three inches. Maximum speed is approximately 270 mph at 21000 feet.

MCDONNELL DOUGLAS Douglas/British Aerospace T-45A (USN). The two-seat intermediate and advanced Goshawk jet trainer was the result of a win in the Navy VTXTS competition of 1981 (over five other candidates) for the British Aerospace Hawk 60. The type was procured to replace T-1C Buckeyes and TA-4J Skyhawks. First flight of the USN revised version was in 1988. The Navy requirement was for over 200 aircraft to be utilized by Training Squadrons VT-21 and 22 at Kingsville, TX, and VT-7 and -19 at Meridian, Ms. with operation started in 1992. Powered by a 5845 pound thrust Rolls Royce Turbomecca F405 turbofan, the 12750 pound gross weight T-45A has a high speed of 625 mph at 8000 feet. Wingspan is 30 feet 10 inches; the Goshawk is carrier compatible and has a single pylon for stores under each wing.

NAVY UTILITY AIRPLANES

Navy utility aircraft have been used, as the name implies, in a variety of missions. In the early days obsolete former tactical types were often used as utility aircraft. Later a few aircraft were specifically designated as utility types right from the start.

One of the first formal VU class aircraft was the UF-1 series starting in 1949 as a production version of the 1947 XJR2F-1 Grumman twin engine amphibian later named the Albatross. The UF-1s were updated in the mid-1950s with a new wing and tail surfaces among other changes to a UF-2 model just as the USAF

SA-16As were modified to SA-16Bs. In 1962 with the major US service aircraft redesignation into a new system, at least for the Navy, the Navy and Coast Guard UF-s were now HU-16Es.

Other Navy utility aircraft were the DeHavilland U-1B Otter of 1956 used in Navy Antarctic exploration work, and a single U-6A Beaver of uncertain lineage obtained from the Army.

A few Piper Aztec twin piston engined light transports were obtained in 1960 for the Reserves as UO-1s in the old designation system with the O being the letter then designating Piper. In 1962 these aircraft were redesignated as U-11As.

DEHAVILLAND U-1B (R.Esposito) A Navy U-1B Otter from DeHavilland Canada is shown at Mustin Field in the Philadelphia Naval Base in July of 1967. Known before the 1962 redesignation as a UC-1, the Otter was first ordered by the Navy in 1956 and these were used in the US "Operation Deepfreeze" Antarctic expedition in 1956-58. Approximately 18 U-1Bs were procured.

DEHAVILLAND NU-1B (R.Besecker) Another Navy Otter, this somewhat modified from standard configuration, is shown in April of 1974 at NATC Patuxent River, Md., this aircraft used as a utility type by the Navy Test Pilot School at that facility. Powered by a 600 horsepower Pratt and Whitney R-1340 Wasp engine driving a Hamilton standard three blade ten foot ten inch diameter propeller, the Otter had normal seating for nine passengers plus two pilots. Wingspan was 58 feet, gross weight as a landplane four tons, and high speed about 160 mph.

DEHAVILLAND U-6A (R.Besecker) A DeHavilland Beaver from the Army is shown in Navy markings. The Beaver was a rare sight in the Navy. Shown in a March 20, 1973 photo at NATC Patuxent River, Md., this U-6A Beaver was powered by a Pratt and Whitney R-985 Wasp Junior engine and had an eight foot six inch diameter Hamilton Standard two blade controllable pitch propeller. Gross weight was 5100 pounds, wingspan was 48 feet, and high speed was about 163 mph.

PIPER UO-1/U-11A (R.Besecker) A Navy Piper Aztec utility aircraft attached to the Reserves unit at NAS Willow Grove Pa. is shown in a photo of July 9, 1966. Originally designated as UO-1s in 1960 when first procured the 20 planes involved were redesignated as U-11s in the major system redesignations of 1962. Additional equipment was provided for the Navy above the stock civil Aztec items.

PIPER UO-1/U-11A (J.Weathers) A photo taken at the New Orleans Lakefront Airport on June 15, 1977 shows another Navy U-11A Aztec twin engined utility airplane and illustrates the longevity of these types. Using two Lycoming O-540 horizontally opposed engines each of 250 horsepower, the U-11As had a normal gross weight of 4800 pounds and a high speed of about 215 mph. Wingspan was 37 feet.

GRUMMAN UF-1 (R.Stuckey) The Grumman Albatross amphibian shown was the UF-1 in the Navy and Coast Guard, and was the equivalent of the USAF SA-16A amphibian. The aircraft started out as an XJR2F-1 in 1947 and later became a UF-1 in production when the JR category was cancelled. They would later undergo extensive revision to UF-2/SA-16B standard. Powered by two Wright R-1820 Cyclones of 1425 horsepower each and carrying a crew of four to six the UF-1 had an 80 foot wingspan, a gross weight of 27000 pounds, and a high speed of just under 250 mph at sea level.

GRUMMAN UF-2/HU-16E (J.Weathers) One of several dozen US Coast Guard Albatross aircraft known as HU-16Es after extensive modification. Wingspan was increased by 16 feet six inches, tail surfaces were increased in area, wing slots were replaced by a cambered leading edge, and various aircraft systems were updated. The prototype flew first in January of 1956, and deliveries began the next year. Some HU-16Es were SA-16Bs obtained from the Air Force.

GRUMMAN HU-16E (USCG) Flight photo of a US Coast Guard HU-16E Albatross shows the amphibian configuration with nose radome and underwing external fuel tanks. The HU-16 aircraft were the general equivalent of USAF SA-16B amphibians. Wingspan was 96 feet eight inches. Power was supplied by two Wright R-1820 Cyclones like the earlier UF-1 aircraft. Empty weight increased about 2780 pounds and gross weight rose to 35700 pounds. High speed was 235 mph and normal cruising speed about 150 mph.

NAVY WARNING AIRPLANES

Replacing the function of the old Navy land or tender based patrol airplanes with observers using the human eyeball and the older carrier based and combat ship based scouting planes, the land and ship based VW category warning aircraft provide the eyes of fleet units by adding the capabilities of long range radar to the cruising radius ability of the AEW (airborne early warning) aircraft. Long range warning of potential enemy movements or weapons launches is vitally important to the Navy. The VU category of Navy planes started in the mid-1950s, but Navy warning planes were integrated into the new 1962 redesignation system.

The first Navy warning airplane was the Lockheed radar equipped Constellation land based AEW type of 1954, first as the PO-1W patrol plane with a W suffix letter for the warning mission and later a WV-1. Changing from the civil 749 to the 1049 Connie

brought out the WV-2 with increased capabilities. These later aircraft became EC-121Ks in the new 1962 system.

The first carrier based AEW airplane for the Navy came from Grumman in 1958 as the WF-2 Tracer, a major modification of the S2F- twin piston engined Tracker aircraft with a big search radar overhead. This type was utilized into the mid-1960s as a standard carrier airplane. In 1962 production was started on a new design twin turboprop powered carrier based AEW aircraft with greatly increased capability of both the aircraft and its avionics. The new plane, still used today as the carrier based AEW type, was the Grumman W2F-1 Hawkeye first flown in late 1960 and shortly in 1962 redesignated to an E-2A, and then delivered also in updated form starting in 1971 as the E-2C. This aircraft and further updated versions will serve the Navy for the forseeable future.

GRUMMAN WF-2/E-1B (J. Weathers) The first carrier-based Airborne Early Warning aircraft employed by the Navy, the Grumman WF-2 Tracer, later in 1962 redesignated as an E-1B, was a considerable modification of the Grumman S-2 Tracker ASW airplane having a very large antenna with a radome atop the fuselage designed by Hazeltine Electronics Corp. An E-1B of Squadron VAW-88 is shown with wings folded at New Orleans International Airport on November 26, 1972. The aerodynamic prototype for this design flew first on December 17, 1956. A total of 88 E-1Bs were produced.

GRUMMAN WF-2/E-1B (Mfr.) Flight view of two WF-2 carrier-based AEW airplanes shows the general configuration emphasizing the features of a large radome requiring twin tails. The 88 production Tracers were manufactured between February, 1958 and September, 1961. Powered by two 1525 horsepower Wright R-1820 Cyclones driving 11 foot diameter Hamilton Standard propellers, the loaded weight of the E-1B was 26600 pounds. Wingspan was 72 feet five inches and length 45 feet four inches. High speed was approximately 225 mph at 4000 feet.

LOCKHEED WV-2/EC-121K (AAHS) The successor to the PO-1W/WV-1 Navy version of the earlier 749 Constellation, the WV-2 was the 1954 land-based Airborne Early Warning (AEW) Navy version of the civil 1049 Connie using about five and one half tons of radar and other electronic gear. The graceful Connie lines were interrupted by an upper fuselage seven foot high radome for a General Electric height-finding radar and a belly radome for G.E. surveillance/distance finding radar as shown in the photo. Initially a WV-2, the type was redesignated in 1962 to an EC-121K aircraft.

LOCKHEED WV-2/EC-121K (USN) A May 18, 1959 photo shows a flight view of a WV-2 Connie airplane named a Warning Star in the vicinity of Diamond Head, Hawaii. Provision was made for a crew of up to 31, including relief pilots, radar operators, technicians, and maintenance specialists. Five radar consoles and plotting tables were provided along with auxiliary radar units. A Combat Information Center (CIC) coordinated all search operations. Powered by four 3250 horsepower Wright R-3350 Turbo Compound engines, the 144000 pound gross weight WV-2 had a wingspan of 126 feet and a high speed of about 320 mph at medium altitude.

LOCKHEED NEC-121K (R.Besecker) A specially modified (N) Warning Star of Squadron VX-8 is shown at NATC Patuxent River,Md. on July 15,1967. Noteable is the absence of the height finding radar normally atop the fuselage; the lower radome has been retained. The aircraft was utilized for a special project and was not modified back to standard EC-121K configuration.

GRUMMAN E-2C (USN) One of two prototypes of the E-2C Hawkeye twin turboprop carrier based AEW airplane is shown. The Hawkeye started off when first flown in October of 1960 as a W2F-1 which was redesignated to E-2A in 1962. Some 56 E-2As were procured between that time and February, 1967 powered by Allison T56 turboprop engines each of 4050 equivalent shaft horsepower driving four bladed 13 foot six inch diameter Aeroproducts propellers. Characterized by the large rotodome over the fuselage and a special ATDS (Airborne Tactical Data System) most all E-2As were updated in 1969-70 with new avionics including a new microelectronic computer. The E-2C, of which 166 were delivered starting in mid-1971 after a first flight in January had uprated 4910 ESHP T56 engines and a new generation of electronics.

GRUMMAN E-2C (USN) Flight view of an E-2C Hawkeye from Squadron VAW-114 aboard carrier USS Kittihawk shows the configuration of the current carrier based AEW airplane with the AN/APA-171 rotodome and AN/APS-120 search radar along with a large amount of other avionics. Wingspan is 80 feet seven inches and rotodome diameter about 24 feet. The two T56 engines provide the E-2C with a high speed of 374 mph and a cruising speed of 310 mph. Maximum takeoff gross weight is 51570 pounds.

NAVY RESEARCH AIRPLANES

The Navy has participated in many aircraft research projects over the years, sometimes alone and sometimes in concert with other services or goverment organizations. Two projects conducted just after World War II are noteworthy.

An early research vehicle appearing in 1947 was the Douglas D-558-I straight wing Skystreak, a jet powered aircraft specially designed to obtain flight data at speeds then beyond those available in wind tunnels. The D-558-I set a 1947 speed record.

A follow-on Navy research aircraft type was the Douglas D-558-II swept wing Skyrocket, this early vehicle using a rocket engine to achieve Mach 2.0 speeds in 1951 and after launch from a mother aircraft able to attain altitudes of over 80000 feet. One D-558-II now resides in a place of honor at the National Air and Space Museum in Washington, DC.

DOUGLAS D-558-1 (UNK.) Conceived in 1945 to obtain research on free flight air load measurements not available from then-current wind tunnels, the single seat Douglas D-558-1 Skystreak was a straight wing jet powered plane with the first of three having an initial flight on May 28, 1947. The D-558-1 set up a world speed record of almost 651 mph on August 25, 1947. The entire nose section was jettisonable in an emergency. The fuselage had 400 pressure measurement points connected to an automatic recording system along with strain gages.

DOUGLAS D-558-1 (Mfr.) Flight view of the D-558-1 shows the small cockpit enclosure and the thin 10% thickness ratio wing. The entire rear section was a thick wall magnesium alloy tube. The engine was an Allison J35 of 5000 pounds of thrust providing the 651 mph high speed. On the speed record flights the D-558-1 takeoff weight was five tons or more and about 230 gallons of fuel was carried along with 640 pounds of research equipment.

DOUGLAS D-558-2 (R.Besecker) Ground view of the D-558-2 Skyrocket shows clean lines and a tricycle landing gear. An entirely different plane from the D-558-1, the three -2s were used for investigation of 35 degree swept wings and had a first flight on February 4, 1948. Powered by both a Westinghouse J34 turbojet engine of 3000 pounds thrust and a Reaction Motors XLR-8 bipropellant rocket motor of 16000 pounds thrust, the aircraft had a span of 25 feet and a high speed of Mach 2.0. With the jet engine removed one D-558-2 attained a high speed of 1238 mph in August of 1951. The rocket plane was dropped from a "mother" aircraft at about 32000 feet. On August 31, 1953 a Skyrocket climbed to over 83000 feet. The planes made a total of 161 flights of which 74 were combined jet and rocket powered.

A-6 catapulting off the deck of USS Nimitz in the Atlantic, September 1980. (USN)

F-4 Phantom ready for catapulting aboard the USS Ranger. (USN)

A-7E Corsair II aboard the USS Midway. (USN)

A-7E Corsair II's from the USS Constellation, November 1974. (USN)

Grumman F-14A Tomcat coming aboard an aircraft carrier. (USN)

Grumman F-14A Tomcat. (USN)